城市设计演变

CITY + DESIGN + EVOLUTION

—— 来自六大洲的特色景观设计

—Unique Landscape Designs from 6 Continents

Pedro F Marcelino《景观设计》杂志社 编著

大连理工大学出版社

图书在版编目（CIP）数据

城市、设计、演变：来自六大洲的特色景观设计：
汉英对照 /《景观设计》杂志社著. -- 大连：大连理
工大学出版社, 2011.3

Ⅰ. ①城… Ⅱ. ①景… Ⅲ. ①景观—园林设计—作品
集—世界—现代 Ⅳ. ①TU986.2

中国版本图书馆CIP数据核字(2011)第021263号

出版发行：大连理工大学出版社
　　　　（地址：大连市软件园路80号 邮编：116023）
印　　　刷：利丰雅高印刷（深圳）有限公司
幅面尺寸：230mm × 290mm
印　　张：22.5
字　　数：405千字
出版时间：2011年3月第1版
印刷时间：2011年3月第1次印刷
策划编辑：苗慧珠
海外编辑：Pedro F Marcelino
责任编辑：刘晓晶
责任校对：万莉立 刘　姝
版式设计：王　江 赵安康

ISBN 978-7-5611-6032-9
定　　价：338.00元

电　话：0411-84708842
传　真：0411-84701466
邮　购：0411-84708943
E-mail:dutp@dutp.cn
http://www.landscapedesign.net.cn

Pang Wei

Landscape is Motherland, is Ourselves

According to Kevin lynch, design is possible formalization by imaginative creation, meeting mankind demands, included social, ecological, aesthetical and technological. Unscrambling it, design is achieving objectives, being to solve out problems, moreover, we ask for imagination, formalization, creativity! Landscape architect design on the land, for the exterior space, however, landscape transformation and social alteration are not grasped by designer's hands but by times itself. Designers paint on canvas unfolded by times as pawns. In all conscious, extraordinary designer could be the legend of era by due.

Regard to pathology, the world we living is abnormal field, pathogeny form antiquity like earthquake and disease, pirate and greed of "the merchant of Venice", are the phenomenon we processing. Simultaneously, globalization and urbanism, development and environment protection paradoxically take the procession of priority problem we facing, landscape design study under the obligation how much effort can ever make?

Landscape architect adhered to the market for making a living, the force of commitment in spite of authority or fortune is potency, and design for potency to portray pageantry or balance droit and potency, moreover to tend to represent droit, the circumstance is confront by designers. The droit required by normal people, by disadvantage groups, by history and future, so much as animal, foliage, steams and valley.

As Chinese proverb said "servant body, miss heart", by the tune of clerisy landscape protest ambition for the land and culture. Landscape architect design for making a living, for animadverting, for creating, being proud of composite occupation.

Who is constructing landscape, will be the marvelous scenery by due.

From our insight, landscape is our motherland, is ourselves.

Night of April 25th, 2009 in Guangzhou

Prizes and Honors He Has Won Include

Honorable Design Prize (top prize granted by ASLA in 2002) with Qijiang Park Project in Zhongshan City, Guangdong Province as the principal designer; Gold Prize of Environmental Art in the 10th National Fine Arts Exhibition and Gold Prize of Modern Excellent Folk Architecture in China (cooperative project) in 2004; Chinese architectural art prize in 2003; Gold Prize Modern Excellent Folk Architecture in China in 2004; Top Honorable Design Prize in the 22th International Urban Waterfront Outstanding Design in 2008; Asia Pacific Outstanding Honorable Prize granted by Urban Land Institute(ULI) in 2009.

Prizes with Baiyun International Conference Center Project (completed in 2006) as the principal designer include: Public Architectural Design Prize at the Barcelona World Architecture Festival in 2008; The 8th China Civil Engineering Zhantianyou Prize in 2008; Excellent Prize of Guangdong province in 2008; Title of "2005-2006 Excellent Landscape Designer of Chinese Real Estate" in 2006.

"Gold & Diamond Prize" of Chinese mainstream estate and title of "The Most Influential Architectural & Landscape Designer of the Year" in 2006; Title of The Best Landscape Designer in Chinese Estate for Three Consecutive Years from 2007 to 2009.

First Prize and the first bid winner in the Planning & Architectural Design Project of Automobile Town in Nanhai District, Foshan City, Guangdong Province.

Superior Prize and bid winner in the Landscape Planning Design for Sanshui Southwest Residential Group Central Area Public Green Space, Foshan, Guangdong Province.

Superior Prize in the International Contest of Planning Design for Huangqi Mountain Urban Park, Dongguan, Guangdong Province.

景观是大地，也是我们自己

借凯文·林奇的话"设计是想像地创造某种可能的形式来满足人们的某种目的，这些目的包括社会的、经济的、审美的和技术的"。解读之，设计是有目的的，要解决人们的问题，并且要有想像、有形式、有创造！景观设计师设计土地，设计户外空间，但改变大地和社会面貌的根本力量并不在设计师，而是大时代本身。设计师是大时代的画笔、大时代的走卒；当然，设计师中的佼佼者，本身就可能成为大时代的传奇之一。

我们今天生活的世界本身就是一个问题场域，古老的地质灾害和疾病、古老的海盗和"威尼斯商人"式的贪婪，都是真实世界的现状。与此同时，全球化与城市化、经济发展与生态保护等等不古老的问题又形势逼人，景观设计学在其中能有几分承担，又有几分无奈？

景观设计师是自谋生计的商品经济依附者，设计委托的力量不管其为行政的力量抑或财富的力量，都是权力之力；而设计是为权力描绘华服还是在权力—权利之间寻找平衡和化解，乃至做出倾向权利的表达，这都是中外设计师面临的境遇。这权利包括普通人的、弱势者的，也包括历史的权利、明天的权利，甚至动物、植物的权利，河流山谷的权利。

用旧时代的话调侃，这真是"丫环身，小姐心"，景观设计师用公共知识分子的口吻说出和做出对山河、文化的抱负，着实让人恼耶？怜耶？糊口生存，批判建树，创造耕耘，景观设计师三位一体，幸甚至哉！

营建风景的人，本身将会成为风景。

我们会看到，景观是大地，也是我们自己！

2009 年 4 月 25 日夜羊城

北京土人景观与建筑规划设计研究院	副院长
广州土人景观顾问有限公司	总经理 / 首席设计师
北京大学景观设计学研究院	客座研究员
广州美术学院设计学院	客座教授
台湾国立勤益科技大学绿色生活科技整合研究中心	咨询委员
吉林省延边朝鲜族自治州	经济社会发展顾问
广东省环境艺术设计行业协会	副会长

Beijing Turen Landscape & Architectural Planning Academy	Subdean
Guangzhou Turen Landscape Planning Co., Ltd.	General Manager / Chief Designer
Landscape Design Academy of Peking University	Visiting Researcher
Design School, Guangzhou Academy of Fine Arts	Visiting Professor
Research Center of Green Life Technology integration, TaiWan	
National Chin-Yi University of Science and Technology	Consultant Member
Yanbian Economic and Social Development	Consultant
Guangdong Environment Design Institute	Vice-president

重要奖项列举

· 2002 年作为主要设计人完成的广东省中山市岐江公园项目，荣获美国景观设计师协会（ASLA）2002 年度最高奖项——荣誉设计奖；2004 年荣获第十届全国美术作品展览环境艺术类金奖及中国现代优秀民族建筑综合金奖（合作）；2003 年中国建筑艺术奖；2004 年中国现代优秀民族建筑综合金奖；2008 年，荣获第 22 届世界城市滨水杰出设计"最高荣誉奖"；2009 年，荣获国际城市土地学会（Urban Land Institute 简称 ULI）2009 年度 ULI 亚太区杰出荣誉大奖。

· 2006 年作为主要设计人完成的广州白云国际会议中心项目，荣获 2008 年巴塞罗那世界建筑节公共建筑设计大奖、2008 年第八届中国土木工程詹天佑大奖、2008 年广东省优秀设计奖。

· 2006 年被评为"2005 年～2006 年中国房地产优秀景观设计师"。

· 2006 年中国主流地产"金钻奖"、最具影响力建筑景观设计年度名人。

· 2007 年中国地产最佳景观设计师。

· 2008 年中国地产最佳景观设计师。

· 2009 年中国地产最佳景观设计师。

· 广东省佛山市南海区汽车城规划及建筑设计（获一等奖并中标）。

· 广东省佛山市三水西南组团中心区公共绿地景观规划设计国际竞赛（获优胜奖并中标）。

· 广东省东莞市黄旗山城市公园规划设计国际竞赛（获优胜奖）。

Kong Xiang Wei

Crisis and Landscape Architecture

In the recent decades, prosperous economic development enriches the material life of people at the same time it also improves the appearance of landscape architecture. With regarding to the China's condition, landscape design in the last 30 years has been developing at a dramatic speed. For example, in a macroscopic view, earth and river changed; in a microscopic view, life and living space changed. In terms of groups, we intend to make changes and to rebuilt and establish new cities and earth landscape. In terms of each single person, we are relying too much on materials support even our mental life is based on the material live. Cities and earth is the place stuffing with materials more than a poetic place where people could relax. Maximum tendency controls landscape as well as our life at the same time.

How is the relationship between the crisis and landscape architecture? In the context of the special year 2009, crisis is undoubtedly the pronoun for economic crisis. In general, crisis includes economic crisis, environmental crisis, biological crisis, and resource crisis. Apart from these, it also includes crisis of faith, race crisis and national crisis. Among these, it is economic crisis influences our life most and even crucial for our survival. On the other hand, biological crisis and environmental crisis closely related with landscape influences our life not that much to devastate our life so far. People feel deposits shrinking and stocks evaporation strongly but they would not pay the same attention on pollution of a river where they cross everyday. Human civilization has not so developed and people are still thinking from their narrow perspective.

We are doubted if the economic crisis would slow down the speed of the ever changing landscape and ease the environmental and biological crisis since the slowdown economic development. Economic crisis can change people's valuation system? The valuation system would shift from materialism and money worship to spirit pursuit? From the experience of Great Depression in 1930s of the U.S., it is likely that the crisis in 2009 would not change anything and we are all trying to recover from the crisis and be much stronger after it.

Slowdown economic development seems like good news to the speedy landscape design and construction which is full of mistakes. But in a word, it is the outcome by our design and will. We could have established a poetic and sustainable home on the single planet we could live. It needs awful respect and love, scientific method and prudent attitude, and a slowdown development.

April 26th, 2009

北京观筑景观规划设计院. 首席设计师. 总经理.
曾任《城市环境设计》杂志执行主编. 北京大学《景观设计学》杂志执行主编.
长期从事景观设计评论及理论研究。

Xiangwei Kong, founding principal, Beijing Guanzhu Landscape & Plandesign
Institute, was the founding executive editor for Urban Space Design magazine
and Landscape Architecture China magazine, published by Peking University. He
is an expert on landscape critics and theoretical research.

危机与景观

2009 年，一场席卷全球的经济危机让整个世界的情绪低落下来。这又会给我们所生活的这个星球的景观带来怎样的影响呢？

多年的经济晴好给人们的物质生活带来了极大的丰富，也促生了这个星球景观面貌的改变。就中国而言，三十年间景观以爆炸的速度发生变化，从宏观——土地江河的变迁，到微观——家居生活的改变。就群体而言，患上了改变妄想症，试图不断更新和建立新的城市及大地景观；就个体而言，患上了物质依赖症，将精神寄托在丰富的物质生活基础上，城市和大地已不再是充满诗意的精神家园，而是物质堆砌的场地。极多主义控制了景观，同时也控制了人们的生活。

如何定义危机与景观之间的关系？在 2009 年这个特定的时间语境中，危机无疑是经济危机的代名词。从人类尺度来讲，危机包括经济危机、环境危机、生态危机、资源危机，当然还有信仰危机、民族危机、国家危机等。除民族与国家存亡之外，经济危机会快速影响到每个人的生活，甚至是生存。而与景观密切相关的生态危机、环境危机反映到每个人的生活中则是一个缓慢的过程，在人们的潜意识中那是群体的事情。一条每日经过的河流受到污染，反映到个人身上远远没有银行存款的缩水和股票的蒸发给人的感觉强烈。这说明人类文明的进程并没有改变人们以自我为中心作为思维与意志的出发点。

经济危机是否会让快速变化的景观减速，现在还不得而知；是否会因经济脚步的放慢而暂时缓解环境与生态危机，减少对资源的占用，也不敢妄下断论。经济危机会改变现代人的价值观吗？集体价值观会从拜物主义和拜金主义转向对精神世界的寻求吗？从美国 20 世纪 30 年代的经济大萧条到 2009 年经济危机爆发的经验来看，似乎不会改变什么，人们梦想的是经济的全面复苏，甚至是复苏后的更强劲。

经济发展的放缓对充满争议的快速景观建设而言，似乎是一个利好的消息。但归根结底，是意志主导的行为对景观构成的影响。营造一个充满诗意和可持续的家园，需要的是一份敬畏和热爱之心、科学的方法和谨慎的态度，以及放缓的过程。

2009 年 4 月 26 日

Huang Zheng Zheng

Deliberation as the Wise — Landscape Industry Status and Its Development

The concern about environment protection and the controversial attitude of cultural consciousness, almost being the keywords midst 2008 top 10 news event of landscape industry. Contingently, the issues came up from the news event is the silhouette of the future trend and the focal point of landscape industry. Landscape not only exterior decoration and gardening, but which concerning about essence of human survival and development. Globalization and urbanism fetched deliberation of environment disaster and self-cognizance suspicion, the challenge imposed more obligations upon landscape design industry.

Our country is catastrophic land as normality, experience from calamity taught our predecessor how to opt for home site, how to use land, utilize water resource. Minority in Yunnan province has cultivated at terrace complied with the contours of mountain, not for visual impact, it is considered to be necessary according to their experience and deliberation. Designer belonging to the wise so as to we should deliberate as the wise. So far, 400 Chinese cities have been short of water supply, 70% surface water is contaminated, and 50% wetland is gone. At this crucial time, the respect of environment is the dominant premise; thus "decoration" is not in ascendance in landscape design industry. The last year disaster—the 5 · 12 Wenchuan earthquake recalled our national-wide united and compassion, simultaneously triggering deliberation and exploration of essence of landscape design, therefore landscape design regressed foundation of land usage and its enthusiasm. As the associated member and organizer of event of "Design after Wenchuan Disaster", GZ TURENSCAPE designed landscape for the schools which collapsed in earthquake. Strategically, safety, economy, and psychology would be the vital consideration for the proposals we provided, such as proposal "vacuums" aroused priority of safety, proposals "color" "game" and "animal therapy" due to the consideration of psychology treatment.

In China, the debate about perception of culture identification was never ceased, beyond it the controversial landscape architectural practiced in very different ways. As the respect of culture, specific village image, local geological landform, and behavior characters of local people dramatically represented culture more sophisticated rather than opening ceremony of 2008 Olympic game held in China. Human being belonging to every pieces of land, for example the contrasted culture between Jiangnan, Beijing and Shudi caused the exclusive language and distinctive food. Esperanto might reveal this fact: Esperanto is universal language created by Dr. Ludwik Lejzer Zamenhof at 1887 in order to help human being use common language to comprehend other races and eliminate hatred, and eventually achieve equality and philanthropism. However, after 122 years spread and practice, Esperanto failed to be the common language. Its failure ascribed to complicated factors included with distinctive language, history and land partition, which the language severed from particular culture, land and groups of people and it definitely lack of vigor. Landscape architecture concentrated on culture distinction is approaching to avoid culture and urban homogeneity

With the regard of every different cultural consciousness, landscape sociology expressly concerned about disadvantaged group, for instance: civilian workers in cities; countrified peasants; factories workers. In the practice of QiJiang park, no other than regard of heritage left by Yuedong dockyard with its land, workers, and industrial specialties which belonged to old times made the revitalized landscape charming and prominent. And more over, philanthropism is shown in plants as well, with strong disapproval and censure of sick aesthetics, standing the opposite side of judgment which partition plants by its market prices. Local plants could be retained by low maintenance to obtain water-saving objective. With accord of contemporary value and aesthetics Landscape industry should provoke healthy, energy saving "colloquialism landscape".

Landscape architects have to deliberate as the wise, whilst design consideration answer for the environment, society context, there were probably no more affairs of TaiHu sea algae, no more drama of Chinese symbolized city. Hopefully, plentitude of comprehension and experience could help China landscape industry grow mature.

April 15th, 2009

Prizes and Honors She Has Won Include

Honorable Design Prize (top prize granted by ASLA in 2002) with Qijiang Park Project in Zhongshan City, Guangdong Province as the principal designer; Gold Prize of Environmental Art in the 10th National Fine Arts Exhibition and Gold Prize of Modern Excellent Folk Architecture in China (cooperative project) in 2004; Chinese architectural art prize in 2003; Gold Prize Modern Excellent Folk Architecture in China in 2004; Top Honorable Design Prize in the 22th International Urban Waterfront Outstanding Design in 2008; Asia Pacific Outstanding Honorable Prize granted by Urban Land Institute(ULI) in 2009.

Superior Prize and bid winner in the Landscape Planning Design for Sanshui Southwest Residential Group Central Area Public Green Space, Foshan, Guangdong Province.

Superior Prize in the International Contest of Planning Design for Huangqi Mountain Urban Park, Dongguan, Guangdong Province.

Coauthor of Landscape · Perspective—Guangzhou Turenscape(2000~2008) Review · Works · Theory (written by Pang Wei, Huang Zhengzheng , Zhang Jian).

国家一级注册建筑师、高级建筑师、高级工程师
广州土人景观顾问有限公司董事、设计总监
2002 年获美国景观设计师协会（ASLA）荣誉设计奖
第十届全国美术作品展览环境艺术类金奖
2009 年中国地产最佳景观设计师

National Grade I Certified Architect and Senior Architect
Guangzhou Turen Landscape Planning Co., Ltd. Trustee, General Director of Design
ASLA Honorable Design Prize, 2002
Gold Prize of Environmental Art in the 10th National Fine Arts Exhibition
Best Landscape Designer in Chinese Estate in 2009;

重要奖项列举

· 2002 年作为主要设计人完成的广东省中山市岐江公园项目，荣获美国景观设计师协会（ASLA）2002 年度最高奖项——荣誉设计奖；2004 年荣获第十届全国美术作品展览环境艺术类金奖及中国现代优秀民族建筑综合金奖（合作）；2003 年中国建筑艺术奖；2004 年中国现代优秀民族建筑综合金奖；2008 年，荣获第 22 届世界城市滨水杰出设计"最高荣誉奖"；2009 年，荣获国际城市土地学会（Urban Land Institute 简称 ULI）2009 年度 ULI 亚太区杰出荣誉大奖。

· 广东省佛山市三水西南组团中心区公共绿地景观规划设计国际竞赛（获优胜奖并中标）。

· 广东省东莞市黄旗山城市公园规划设计国际竞赛（获优胜奖）。

· 2008 年与庞伟、张健合著《景观·观点——广州土人景观（2000～2008）评论·作品·理论》一书。

像智者一样思考——景观行业现状与发展

2008 年度评选出的中国景观行业十大新闻事件中，对环保及生态的关注和对历史文化的态度与争论，几乎成为所有入选新闻事件的关键词。或许，这从侧面反映出当今景观行业未来最应关注的焦点及发展趋势。景观行业已绝不是"涂脂抹粉"的室外装饰或纯粹造园术，它关系着人类生存与发展的本质。全球化与城市化带来了生存环境的生态危机及文化身份的认知危机，让中国的景观设计行业承担更多的重任和更大的挑战。

灾难的经验让我们的祖先学会了如何择地而居，如何使用土地、利用水资源。云南原阳哈尼族人依山而建的梯田，不是为视觉的震撼，而是蕴含着与耕种相关的一切经验与思考。设计师是智者从事的职业，所以需要像智者一样思考，顺"自然之势"而为。目前，中国约有 400 多座城市缺水，70% 的地表水被污染，50% 的湿地消失，如此严峻的生存危机之下，景观设计行业再不可能冒"生态"之大不韪而只是去"粉饰太平"了。2008 年的 5·12 汶川大地震，不仅撼醒了国人久违的爱心和团结一致，同时引发了设计界关于设计本质的思考与探索，景观设计行业也再次重归土地设计与监护的生存技术与艺术的思考。在"2008 年广东设计界汶川地震设计集结号"活动中，广州土人景观作为其中一名中坚的组织者与参与者，从景观的角度为震后学校重建的景观设计出谋划策——从安全、经济、心理等多方面思考，提出了供安全的灾难庇护场所的"空地"方案，供心理治疗的"色彩""游戏""动物疗法"方案，进行了许多有益的景观尝试。

在中国，关于文化身份认知的争论一直不绝于耳，景观设计由此产生的实践误区也一直不绝于目。因此，泛滥中国大地的诸多异域风情景观，或号称"东方威尼斯"的水乡策划案比比皆是。所谓文化，应该不仅是北京奥运会开幕式所表现的典型中国文化，同时还是依存土地上某一具体乡村、某一具体人群、某一具体地理环境上鲜活的地方文化。一方水土养一方人，譬如江南文化与北京文化或蜀地文化的差异，单从饮食及语言上就迥然不同。这或许可从世界语的发展获得启示：世界语是波兰医生柴门霍夫博士于 1887 年创制的一种语言，目的是人类借助这种共同语言，增进民族间相互了解，消除仇恨和战争，并最终实现平等、博爱。但是，历经 122 年的世界语至今也未成为世界通行的语言。究其原因是由于语言与文化、土地、历史等诸多因素有着密切而错综复杂的关系，脱离具体的文化、土地和人群的语言是没有生命力的语言，景观亦然。而景观对文化差异的重视与运用，正是避免文化同质、城市同质的有力途径。

景观要关注不同群体的文化意识，特别是弱势群体的文化，如城中村的农民工文化、乡野的农民文化、厂区的工人文化等。在岐江公园项目实践中，正是这种对发生在旧粤东造船厂那片土地的人、时代和事物的关注，让岐江公园产生了一种与众不同的气质与魅力。对人的普世态度同样适用于植物，反对所谓植物分贵贱、野花野草不入法眼的陋病，抛弃以病树、病梅、病石为美的病态文化，通过使用生态的、低维护的乡土植物品种来达到节水环保目的。"五四新文化运动"曾提倡的"白话文"运动，而景观行业也需要健康、低能耗、符合时代价值观和审美的"白话景观"运动。

设计师要像智者一样思考，当设计能融入正确的设计语境中时，或许中国将不会再出现太湖蓝藻事件，也不会再出现建中华文化标志城的闹剧，或许能恰当地处理申遗成功后保护与开发的关系，中国的景观设计行业也将更加成熟。

2009 年 4 月 15 日

sheng mei

Starting and Growing—Influence from the World on Landscape Architecture in China

Starting from scratch, the profession of landscape architecture has taken only about 10 years to develop into a well recognized discipline in China. Compared with the history of Chinese urbanization, it happened like overnight. The profession itself went through a typical China Model- growing from simply copying and learning ideas from others, to selecting and exploring new ideas.

As a matter of fact, the professional development of landscape architecture in China has been largely influenced by international design corporations, especially those from North America and Europe. Because of their nature, these corporations have to keep looking for new development areas all over the world, and naturally taking the leadership of the professional development in some emerging markets. Sine the middle of the 1990s, designers from foreign countries, and Chinese designers educated overseas, started to come into China. The approaches, philosophies and concepts that they brought in were completely new at that time, and of course, exciting to everyone. This gave a shock to the design field that had been dominated by traditional Chinese gardens for a long time. Partially because of this, Landscape Architecture is more or less an Import in China, representing a more advanced and innovative way of thinking that was expected to deal with problems in urbanization.

The last ten years can be considered as the start of this profession in China. During this time, one of the jobs that Chinese landscape architects do is to collect design ideas from different countries and apply them to local projects. Due to this reason, landscape design in China has been, and will be, highly westernized. That is to say, the principles, approaches and evaluation criteria are based on those developed in the West. Obviously, this does not only happen in this field, but also in all design related professions and many other fields. To make it simple, it is one character of the development stage that China is going through.

So the communication between China and world, in landscape design, was importing to China in the first ten years, learning and copying ideas at the level of forms. China started to apply new techniques gradually, and even to export. The first thing to export is job opportunities. Because of the amount of construction, this country became a gold mind for world designers and companies. At the same time, intensive practice made designers in China grow fast in short time and come up with some unique design and experiences. These low-tech approaches in the eye of the local market became innovative ideas in western countries. From being influenced to being influencing (very little though), the development of landscape architecture profession is getting into new stage.

As we see, innovation and development is sometimes quite simple: if you can solve a unique problem, you find a unique way. When the target is the top international level or creating new style for China, we may not be able to have any innovation. Instead, ideas may come to you when you forget all these ambitions and evaluate yourself honesty and deeply. So when we read this book, we could learn and get more information and ideas on one hand, on the other hand, we will be able to take this chance to evaluate ourselves. By studying the cases here from the world, we could see the diversity of thinking in different culture and social background, as well as all kinds of ideas and approaches coming from them. Although these ideas and principles may be different and even opposite to each other, we can learn to find our answer by trying to compare and select among them. The more we find, the better our standard will be built. This standard will then develop into a value that is shared with others, and even become the foundation of the value of a person, a profession or culture.

1997 年于天津大学建筑系获工学硕士，2002 年于美国伊利诺伊大学（University of Illinois）获景观建筑学硕士（MLA），美国景观建筑师协会会员（ASLA），现任美国 ATA 设计公司丨劳伦斯集团（北京）设计总监。
自 2004 年在中国完成项目：天津万科东丽湖湿地公园景观设计、北京西单文化广场改造景观设计、北京龙湖滟澜山别墅景观设计等。
研究领域：
2003 至今，生态水环境设计与实践
2002 年，日本街道空间研究，美国伊利诺伊大学瑞尔森奖金（Edward L. Ryerson Traveling Fellowship）资助题目
1994 年～1997 年，清代皇家园林的主题构成与审美，天津大学国家自然科学基金项目课题

Ms. Mei Sheng received Master of Engineering from Tianjin University, China in 1997, and Master of Landscape Architecture from University of Illinois at Urbana-Champaign in 2002. She is now the Design Principal of ATA Lawrence Group Beijing office and is the associate member of ASLA.
Ms. Sheng has practiced in China since 2004, the projects she finished include: Tianjin Vanke Dongli Lake Wetland Design, Beijing Xidan Culture Plaza Renovation Landscape Design and Beijing Ginkgo and Rose Garden Residential Landscape Design.
Her research interests are:
Sustainable water environment design, 2003 to present;
The Research of Traditional Japanese Streetscape, 2002, project funded by the Edward L. Ryerson Traveling Fellowship in UIUC.
The Study of Design Theme, Structure and Evaluation in Royal Gardens of Qing Dynasty, 1994-1997, project funded by National Natural Science Fund in Tianjin University.

起步与发展——世界景观对中国景观的影响

在中国，景观设计成为一个独立行业并受到普遍关注，基本上是发生在近十年内；相对于中国的城市化发展历史，就差不多是一夜之间了。从对国外设计不假思索地复制、模仿，到主动选择并开始探索不同的方法，景观设计专业也经历着中国特色的发展历程。

客观地讲，中国景观设计的发展很大程度上受到国外事务所的影响，尤其是欧美等地的跨国设计公司。这些公司因为自身的发展，需要在世界范围内不断地寻找新的增长空间，经常会担当起一个新兴地区专业领路人的角色。在 20 世纪 90 年代中后期，国外设计师和在国外受过教育的中国设计师在当时带来了令人耳目一新的工作方式和让人兴奋的概念，对长久缺少变化的设计领域形成了巨大冲击。正因如此，在中国，"景观"始终有一种舶来品的身份，它似乎暗示着一种更先进、更国际化的思潮与方法，用来应对中国城市化进程中的种种问题。

过去的十多年，应该算是景观专业的起步阶段，此间设计人员主要的任务之一就是大量收集国外资料并用到国内的项目中。这在相当长的一段时间内，促成了中国的景观设计都是（并将是）"泛欧美"化的——从设计手法、工作方式到价值标准，都会以欧美的标准为主要参照。当然，这个现象不是本专业独有，在中国的整个设计领域乃至其他行业都很普及。简单地说，是中国目前发展的一个特征。

中国景观专业的国际交流，前十年基本是以吸收为主，集中体现在形式上，后来出现了一些技术上的引进，并开始向世界输出。首先输出的是工作机会，中国的建设量使这里成为很多国外设计师的淘金之地。同时，大量的实践机会让中国设计师在短时间内迅速成长，并积累出一些独创性的设计和经验。有些东西在中国是"土办法"，到国际上则成了有新意的东西。这种情况虽然不多，但从单纯地被影响到开始产生影响，中国景观的发展也算迈出第一步了。

其实创新和进步有时并不复杂：解决一个独特的问题，就会出现一个独创的方法。当我们迫切想要赶超世界水平、创立中国风格的时候，可能什么也创造不出来。而当你抛开这些雄心壮志，静心审视自己之后，很多东西会自然浮现。所以学习这本书中的设计、开阔眼界是一方面；另一方面，促成对自己的一种审视和反思。从诸多实例中，会看到不同文化与社会背景下的思维方式，以及由其思维方式产生出的各种概念与方法，这些方法可能大相径庭，理念也或许相互对立，但通过比较和研究，会逐渐明确自己的取舍。取舍多了，会形成一个相对固定的标准，进而发展成一种价值观念。这个价值观将成为一个人，甚至是整个行业或文化立足的根本。

目　录
Contents

非洲篇
Africa

亚洲篇
Asia

大洋洲篇
Oceania

欧洲篇
Europe

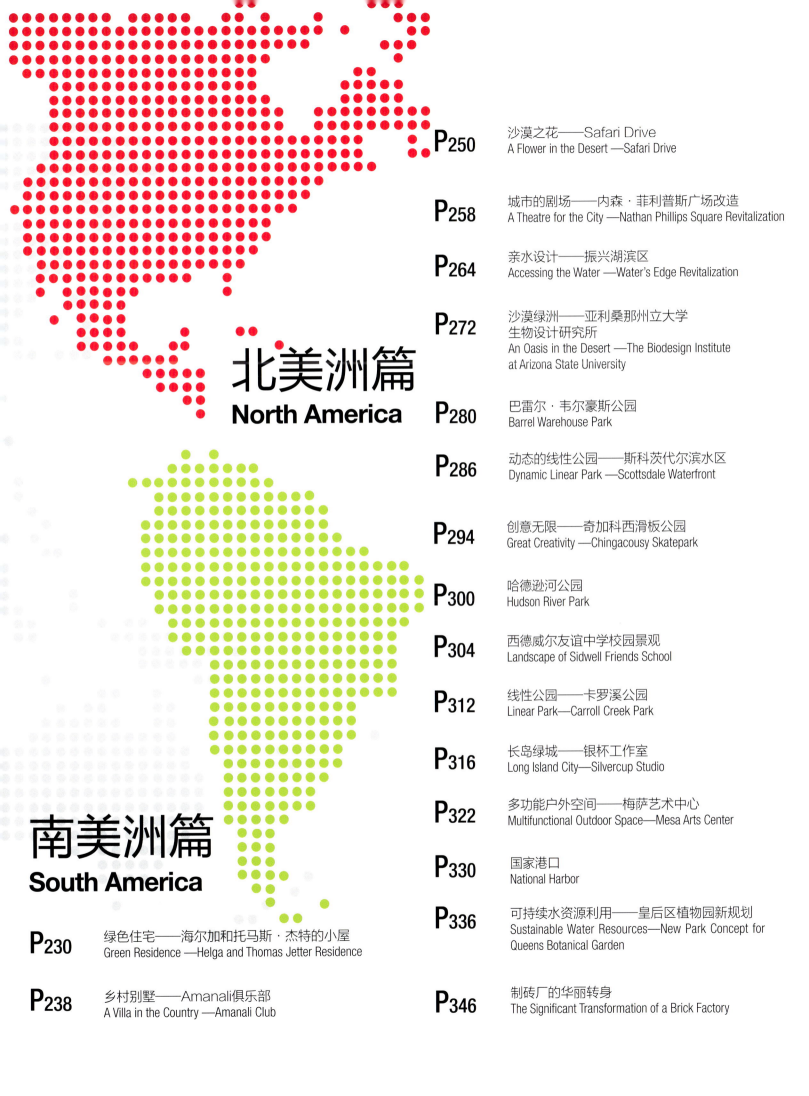

Africa 非洲篇

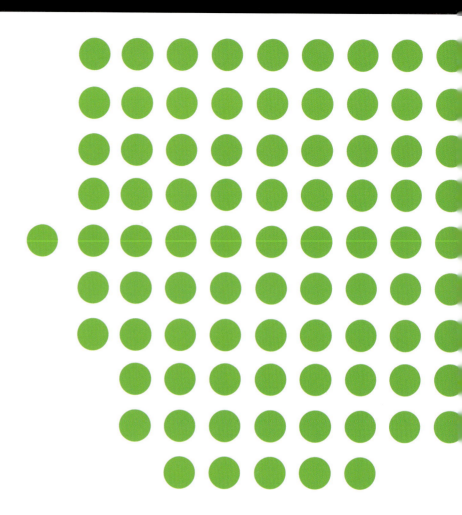

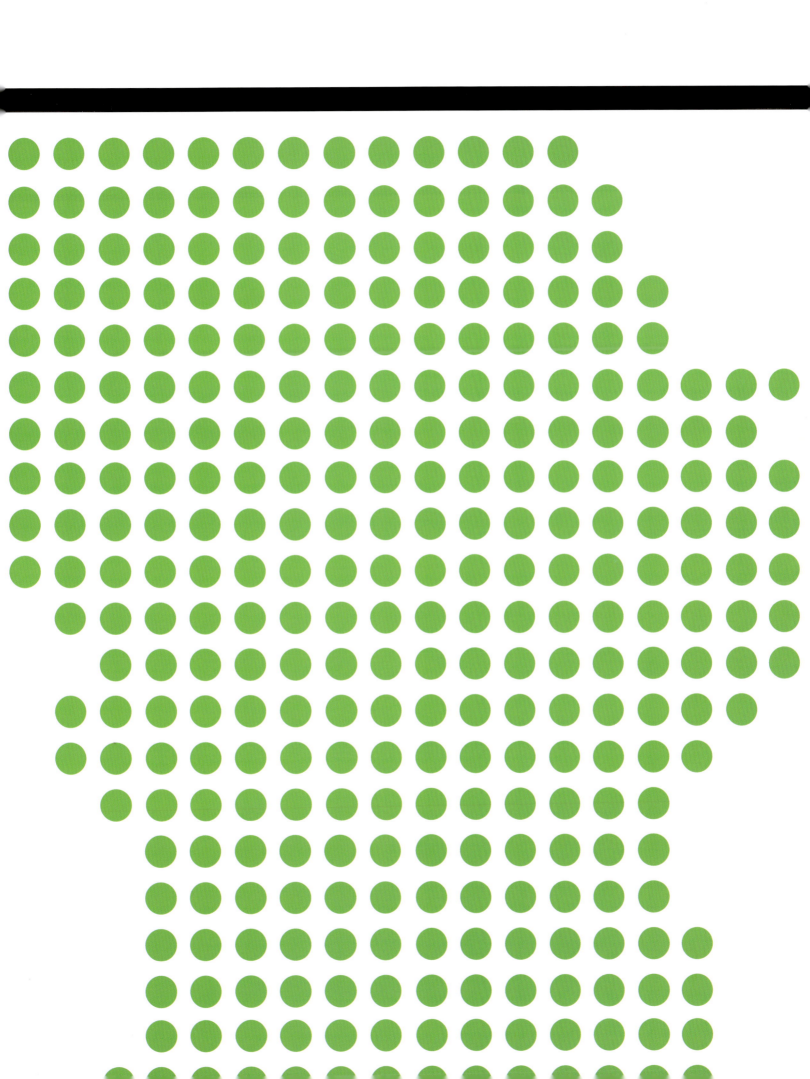

Tanya de Villiers

Principal
CNdV Africa (Cape Town, South Africa)
www.cndv.co.za

Africa is a land of opportunity. There is a saying in the South African development industry, which goes "if you're not doing well, you can only blame yourself". For the last three years we have had a boom in our property industry, which has not been experienced in at least two decades. The boom has meant that landscape architects had the opportunity to get involved in large-scale projects, such as residential estates containing thousands of homes, a large number of leisure developments, golf estates, vineyard estates, olive estates, hotels, resorts and coastal estates. We have opportunities to design at a scale seldom seen in the developed world.

Development in South Africa also has many challenges. There is a definite lack of choice of materials and often a lack of budget for landscaping. Greening of a site is generally seen as a luxury rather than a necessity, but this attitude is changing. The relatively less rosy financial situation that we now find ourselves in has in some way focused the need for higher quality landscapes in order to sell developments. The need to have functional open space is becoming appreciated.

In low-income areas, or government-funded projects, landscaping has to be pared down to the very basics, and needs to be sustainable in the long-term. Public funds are largely channeled into developing, servicing and formalizing "shanty towns". Low maintenance often means hard surfacing combined with trees, as opposed to planting which require maintenance. Trees used in such areas also have to be large and robust enough to withstand vandalism and livestock, notably wandering cattle and goats!

Scarcity of resources and budget in low-income areas force the landscape architect to use their creativity and unconventional materials in their designs. There is an over-arching emphasis on job-creation, so the use of labour-intensive methods of construction and manufacture are essential.

The need for affordable housing has resulted in densities of between 60 to 100 family units per hectare, with the need for high quality useable public open space becoming prioritized in these dense residential areas. The authorities don't have the funds to maintain landscaping, thus developers of residential estates usually have to take on the burden of maintaining landscapes, even in publicly zoned areas.

One of the permeating constraints for projects in South Africa is our low rainfall. Provision of water storage areas, utilization of grey water, groundwater/boreholes and the addition of water-absorbent gels to the soil to help ensure establishment of plants is commonplace. The use of locally occurring – or indigenous – plant species is also commonplace, to ensure that minimal irrigation is required. Existing indigenous plants are collected and potted for re-use by means of "search and rescue" programs prior to developments starting.

Authorities have begun insisting on the use of grey water or cleansed effluent (sewage) water for new developments, whether they are golf courses or residential estates. In numerous cases, the use of clean drinking water to irrigate large-scale landscapes is being prohibited, resulting in kilometers of pipelines are installed by developers to pump effluent water from the nearest sewage works.

Even in this worldwide economic climate, there are numerous exciting projects, where we as landscape architects have the opportunity to create really unique landscapes, in South Africa and further afield.

CNdV Africa总负责人（南非开普敦）
www.cndv.co.za

非洲是一个充满机遇的地方。在南非工业界有这样一种说法，"发展不好怪自己"。近三年，非洲的房地产业繁荣发展，其迅猛之势是近20年来从未有过的。在这样的背景下，景观设计行业有更多参与大型工程的机会，例如：有着数千户居民的小区、高尔夫球场、葡萄园、橄榄树林、宾馆、度假村和滨海区域，这样规模的项目在发达国家都很少有的。

在南非发展也会面临很多挑战。因为这里可供选择的原材料非常有限，景观设计的预算也很少。在此之前当地居民视绿化为奢侈而非必需，目前这种观念正在逐渐改变。由于预算有限，就要注重景观设计的性价比，修建多功能露天广场是最好的选择。

贫困地区的项目或是由政府投资的项目，其景观设计既要经济还要耐用。财政支出主要用于改造旧城区、提升公共服务设施以及城市规范化的建设，而用于后期维护的经费则非常少。由于植被后期所需的维护成本较高，所以利用树木打造景观。设计师选用高大、生命力顽强的树木进行栽植，以抵御各种人为以及牛羊等牲畜的破坏。贫困地区的资源稀缺、经费紧张，这就要求景观设计师充分发挥创造力，打破传统的选材习惯。而在工程中创造更多

的工作岗位是客户对设计师的额外要求，因而设计师要在施工过程中利用当地劳动力密集的特点。

住宅价格需要照顾居民的承受能力，每公顷土地要容纳60户~100户住户，在这样密集的居民区，公共场所就要被高效地利用起来。政府不会在景观的维护上拨款，因此房地产开发商通常要自己承担景观设施以及公共场所的维护费用。

在南非施工的另一个普遍存在的约束条件就是降雨量偏少。到处都在利用蓄水池中的水、再利用的废水、地下水，以及在土壤表面添加吸水剂来确保植被的生长。"搜索与救援"项目早在施工前就已开始，将采集到的本地植被种子种在花盆里以促进其繁衍。

无论是修建高尔夫球场还是居民区，政府都坚持在城市改造中充分利用中水和再回收水。将饮用水用于大型景观的灌溉中通常是被禁止的，因此开发商需要修建几千米长的管道来抽取附近工厂排出的污水。

在全球经济不景气的大环境下，景观设计师还是有机会在南非及其他地区营造出令人振奋的非比寻常的景观。

绿色走廊——的黎波里绿带

Green Corridor—Tripoli Green Belt

翻译　刘建明

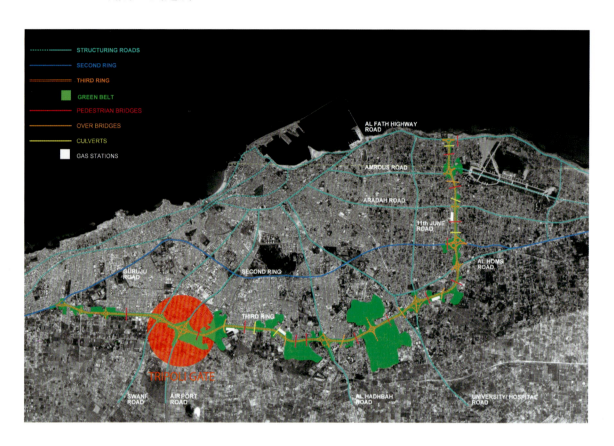

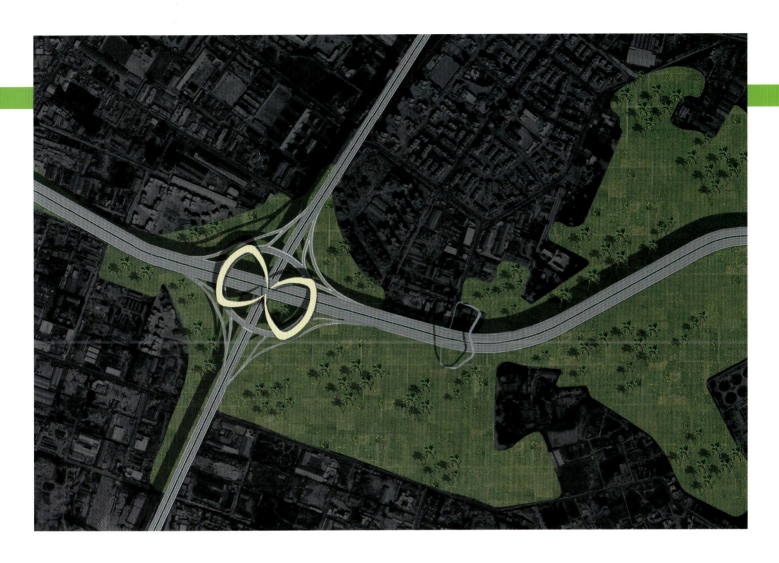

　　的黎波里三环路是一条绵延 24km 的在建机动车道，全路环绕利比亚境内这座新兴的都会城市而建。作为旨在解决日益严重的城市交通拥堵问题，三环路这一重要城市基础设施的兴建又衍生出一些新的问题以及亟待解决的新的城市需求，例如噪声污染、大气污染、无人管理的间隙地带、绿地及公共空间的严重缺失。

　　"的黎波里绿带"试图将三环路整合到城市规划蓝图中去，绿带的规模是的黎波里城市空间规划的一个关键因素。绿色走廊填充在三环路沿线的间隙中，如同流淌的石油。绿带既可以作为阻绝车流噪声和污染的缓冲带，也可作为连接重要地标的纽带。将植物而不是建筑物作为城市化的载体，既可强化公路间隙的定位功能，也能营造出自然渗透的空间效果。

　　为了与本地城市环境配套，本项目配置了部分特色城市基础设施，主要是中心城区没有的各种服务设施。靠近机动车道节点的多种综合服务设施包含温泉旅馆、会议中心、马术中心和城市公园 (Tarabulus 公园)，公园里还将配套体育场、图书馆、非洲研究中心和阿拉伯文化中心。

　　"的黎波里门"是本案的 大特色，它突出了设计中新颖的"向心"特点和地域感。由于机场路是连接中心城区的关键纽带，因此机场路与三环路的交叉处便成为的黎波里醒目的地标，并成为这个城市与世界紧密相连的象征性标志。正是出于这个原因，为了扩展这种文化碰撞的基本内涵，该项目设计了两排从功能上来说迥然不同的对称建筑体——一边是会议中心和办公区，而另一边却是多功能的旅馆和温泉。

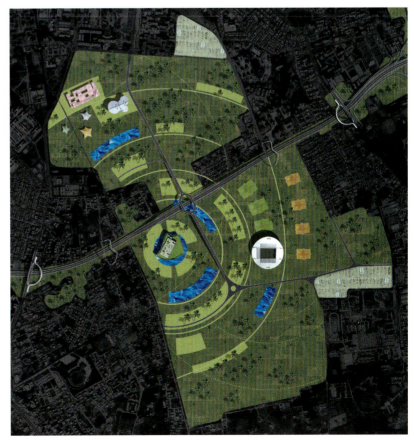

The 3rd Ring Road of Tripoli is a 24km long motorway currently being built around Lybia's booming capital city. As well as new, being an important urban infrastructure, which equips the city for growing car traffic, it creates novel problems and new urban needs that have to be solved, including noise, atmospheric pollution, uncontrolled interstitial zones, and scarcity of green areas and public spaces.

The "Tripoli Green Belt" is an urban project that tries to incorporate the 3rd Ring Road motorway with the urban planning made necessary by the increased traffic capacity. The scale of these implications is a serious issue in the spatial planning of Tripoli. This green corridor along the whole of the 3rd Ring's length fills the interstitials like a mass of oil. It works like a buffer zone against noise and pollution caused by car traffic and connects a sequence of important landmarks. Trees, rather than buildings, will serve as vehicle of urbanization, which will provide the interstitials with a strong identity, producing density with natural permeability.

Responding to local urban circumstances, punctual urban equipments have been proposed, concentrating a wide range of services that were missing in the city. Located close the motorway nodes, these multi-service complexes merge a spa hotel with a congress centre, a horse riding centre and a urban park – Tarabulus Park – in which will be located a sports campus, a library, a centre for African studies and an Arabic cultural centre.

Tripoli Gate is one of the key features of the intervention, and will serve to create a new centrality and sense of place. With the Airport Road as an important connection to the city centre, its intersection with the 3rd Ring Road motorway is a moment of great significance for Tripoli. It becomes a symbolic key connection the city to the world. There was the need to translate this meaning as a significant urban landmark. For this reason, in order to extend the fundamental moment of an encounter of cultures, the project proposes two symmetrical volumes, very distinctive from the functional point of view: one building houses a congress centre and office space, while the other serves different uses, housing a hotel and spa.

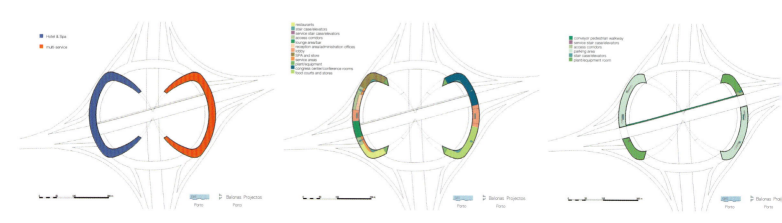

TRIPOLI GATE   PB2   OB 70   PB3

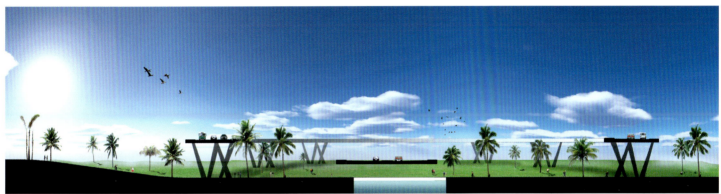

项目位置：利比亚的黎波里

客　　户：利比亚巴西建筑与开发公司

预　　算：6200 万美圆

占地面积：51 760 m²

建成时间：2008 年

建筑设计／景观设计／城市规划：

Balonas Projectos（葡萄牙波尔图）

Location: Tripoli, Lybia

Client: Lybian Brazilian Construction and Development

Budget: US$62 million

Surface Area: 51,760 sqm

Completed Time: 2008

Architects/Landscape Architects/Urban Planners:

Balonas Projectos (Oporto, Portugal)

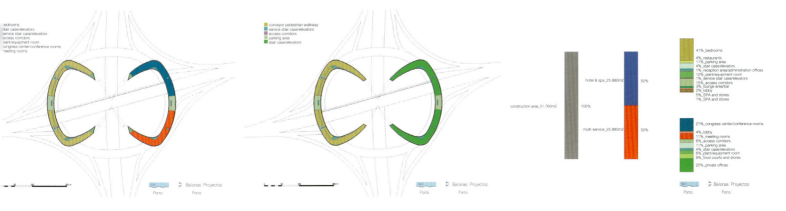

果园中的餐厅

The Orchard Restaurant

翻译　王玲

建筑师最初将餐厅规划为一座位于场地边缘、可以俯瞰法兰斯霍克山谷秀美风景的大房子。然而，这种设计无法满足建筑苛刻的审美需求和视觉冲击。此外，业主希望餐厅的设计能够别出心裁，体现出当地的历史和传统特质。场地原本是一个老旧废弃的、玩滚木球的草坪，它在场地旁边的酒店改造和施工过程中被完全破坏。尽管草坪拥有地表排水系统，但是压实的土壤深层和整个场地需要配备新的地下排水系统，以此来确保新植被的排水要求。

景观设计师提出"果园中餐厅"这一设计理念，不禁令人联想到这里遍布山谷的果园。每位就餐者拥有自己的温室或者凉亭，可以一边赏山谷美景，一边品尝美酒佳酿，完全沉浸在瓜果飘香、芬芳四溢的氛围中。传统的线形沟渠和fine misters组成的水景为炎热的夏季带来了丝丝凉意。

该项目是当地政府在冬末时节批准的，这也就意味着之前已经成熟的树木（如李子、苹果、柠檬和梨）需要在春天来临之前从废弃的老果园中尽快移植过来，这非常棘手。因为，在树木种植的同时其他相应的配套服务设施都必须建立起来，如来自酒店现有的排水管道、卫生间新修的排水管道、雨水排放系统、农用排水渠、温室中的天然气管道，以及景观照明、温室和灌溉所需的能源供给。

项目的各个方面都必须认真协调、井然有序。由景观设计师负责项目的设计管理以及协调工程师、室内设计师和施工人员之间的关系。在修建温室、铺设混凝土板和修建道路的时候，树木周围1m以内都不允许挖沟设渠，而种植槽和铺装材料都是专门定制的。

果树来自于当地成熟的果园，有些来自设有篱架的李子园，有些则是传统果园中的苹果树、梨树或柠檬树。还有30%的果树被放到种植槽中作为储备之用。新种的李树与老李树混杂在一起，它们在白色木架上进行整枝，几年后将取代老李树。

业主希望能够将好望角地区农场典型植被——玫瑰，融入到设计主题中。整个果园采用有机耕种，以确保各种水果和香草随时都可以食用。

室内外空间彼此交融，和谐共生。当夜幕降临之际，果树上灯光璀璨，为这种内外交融的景观体验平添了几分浪漫的情调。

The original architect's plan for the restaurant consisted of a large building, perched on the edge of the site, overlooking the Franschhoek Valley. There was a concern that it would not meet the stringent aesthetic and low visual impact requirements of buildings in the visually sensitive Franschhoek Valley. The client also felt that the restaurant had to be unique and had to relate to the history and traditions of the area. The site was an old, disused bowling green and was completely destroyed during renovations and building operations for the hotel alongside it. Although the green had upper surface drainage, the lower layers consisted of compacted fill, and the entire site had to be fitted with underground drainage to ensure good drainage for new plantings.

The idea of a restaurant within an orchard was proposed by the landscape architect, reminiscent of the numerous fruit orchards typical of the valley. Each group of diners would have their own greenhouse or pergola in the orchard, set amongst fruit trees, roses and herbs, with views over the valley, while chefs cooked up a delectable buffet. Water features consisting of traditional "grachts" (or linear canals) and fine misters were added to cool the space on hot summer days.

Local authorities approved the concept in late winter, meaning old mature trees (plums, apples, lemons and pears) had to be sourced urgently from disused or old orchards and transplanted before spring. This was a logistical nightmare, as trees were being planted at the same time that a myriad of services (namely existing sewer lines from the hotel, new sewers for the toilet facilities, stormwater drainage, agricultural drains, gas lines to all the conservatories, power to landscape lighting and conservatories as well as irrigation lines) had to be accommodated and installed.

Services therefore had to be carefully coordinated and the project proved very challenging as no trenching was allowed within 1m of any tree while contractors were installing greenhouses, concrete slabs and pathways. Pots and paving materials were custom designed for the project. The landscape architect was responsible for the design, project management and coordination of the engineers, interior designers and builders.

Orchard trees were sourced locally, from mature orchards in the valley, some from espaliered plum orchards, and apples, pears and lemons from conventional orchards. An additional 30% were put into containers as reserves. New plum trees were inter-planted with older ones and will be trained onto white-washed timber frames to replace the old trees in years to come.

The client requested that roses, traditionally used on Cape region farms, be incorporated as part of the theme. These were combined with edible herbs and fragrant hedges. The entire orchard is farmed organically to ensure that all fruit and herbs are edible at all times.

Interior and exterior spaces are now almost indistinguishable from each other, during the evening trees are lit to create a wonderfully romantic setting for this popular indoor-outdoor landscape experience.

项目位置：南非开普敦法兰斯霍克
占地面积：2500m²
预　　算：52 万美圆
项目时间：2006 年~ 2007 年
景观设计：CNdV Africa 景观设计事务所的 Tanya de Villiers
客　　户：3 Cities Hotel Group
室内设计师：Beth Murray
园 艺 师：Meg Pittaway
所获奖项：2007 年南非景观设计师协会金奖

Location: Franschhoek, Cape Province, South Africa

Area: 2,500m²

Budget: US$520,000

Project Dates: 2006 – 2007

Designer/Company: Tanya de Villiers at CNdV Africa

Client: 3 Cities Hotel Group

Interiors: Beth Murray

Planting Implemented By: Meg Pittaway

Awards: South African Landscape Institute Gold Award, 2007

Asia 亚洲篇

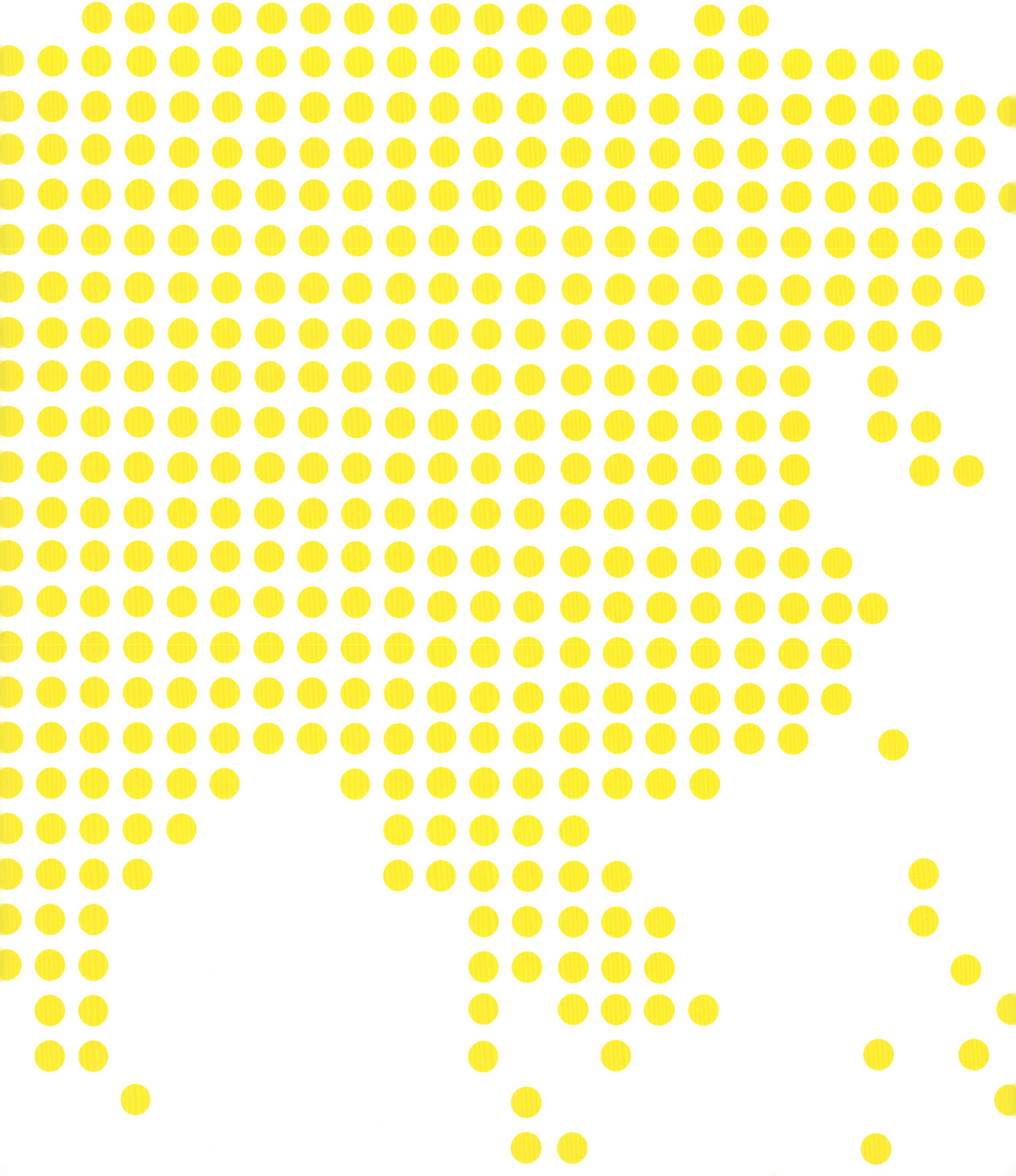

Vladimir Djurovic

Principal
Vladimir Djurovic Landscape Architecture (Broumana, Lebanon)
www.vladimirdjurovic.com/

Blessed by an incredible economic growth, these distinct regions have witnessed tremendous increase in new developments and projects throughout, at an astounding rate. Even though the field of landscape architecture as such is relatively new in these regions, the attention to it and its spread, especially towards the end of this decade was phenomenal.

Whether commercial, institutional, public, or private, the emphasis placed on landscape architecture was surprisingly remarkable. Landscape architecture was considered in many instances as the single most important component to differentiate and identify a project, and in some cases affected the overall success or failure of these projects.

Hence, this has opened tremendous opportunities for exploration, and experimentation in this field, but most importantly it has given landscape architects a unique chance to solidify the real role, and real value of what proper landscape architecture can offer to communities and to life as a whole.

Underneath such momentum, and certainly at a global level as well, there is a brewing need to reconnect to something real, something pristine, and something natural. Hence, in this fast-paced era, nature, and bringing people closer to it, is a fundamental necessity that cannot be overlooked. It's precisely that need that is offering meaningful opportunities to pursue in these two distinct regions.

Asia is a bewildering land, blessed by awe-inspiring natural beauty, biodiversity, and infinitely rich ecosystems, and a climate conductive to for outdoor living. Flourishing as the escape destination of the world, the main challenge for landscape architecture in Asia would be to preserve and seamlessly integrate into its environment and culture.

Landscape architecture should have the absolute minimal impact when trying to meet its requirements, while offering itself almost as a tray to appreciate and enjoy this magnificent natural environment.

Equally stunning, yet totally opposing, is the Middle East. In its majority, it is a more arid and dry region, with an exceptionally harsh climate. An "exotic" destination in its own sense, it is being pursued as the new region to be discovered. A region where opportunities are great, but challenges for landscape architecture are even greater. Even though the Middle East was the birthplace of some of the most inspirational gardens in the world, most of its projects today, a few thousand years later, are mere replications of gardens in the western world. There is a serious urge to reinvent these lost gardens of paradise, through landscapes that induce enjoyment of the outdoor in an exemplary and responsible way.

As growth and development will undoubtedly continue to soar in both regions, landscape architects should seize every opportunity to create environments that reflect their context and age, become models of respect to nature, and reassume their role of being an inspiration to reconnect to the universal.

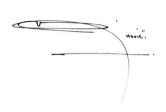

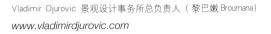

Vladimir Djurovic 景观设计事务所总负责人（黎巴嫩 Broumana）
www.vladimirdjurovic.com

随着经济的快速发展，独具特色的亚洲及中东地区的规划和发展也突飞猛进。尽管景观行业在该区域新兴不久，但近十年来已备受关注，其景观设计风格也流行于各地。

无论是商业区还是其他用地、公共空间亦或是私人空间，景观设计都显得尤为重要。景观设计一直被视为标新立异的核心元素，甚至是决定整个项目成败的关键因素。正因如此，景观设计这个领域有着巨大的探索和发展空间。更重要的是，设计师进行创作时，其自身价值和项目的价值均得到了最大化的体现。

从全球发展的角度看，在这样快节奏的时代，景观设计中迫切需要一些真实、淳朴、自然的元素。因此，与大自然亲密的接触显得十分重要。在亚洲和中东地区这种需求更加迫切。

山河壮丽、物种富饶、生态系统完善的亚洲是一个迷人的地方，气候非常适宜户外活动，也是各地人们钟爱的度假胜地。因此，其景观设计的最大挑战在于如何在不破坏原有环境和文化的前提下，巧妙地加入新元素。景观设计师在满足设计需求的同时要尊重自然、融入自然，并尽量减少对环境的影响。

然而，中东地区的情况却与之恰恰相反——中东的大部分地区干燥荒芜、气候环境恶劣，这里充满异域风情，是一片亟待开发的土地。虽然这里的机会无限，但对于景观设计师而言挑战也更大。尽管历史上很多杰出的花园设计都出于中东地区，但几千年以后的今天，这里的大多数工程都是照搬西方国家的设计理念。中东地区急需通过景观设计来重现往日的辉煌，给人们的户外活动增添更多的欢乐。

毫无疑问，亚洲和中东地区的发展形势大好，景观设计师应该抓住机会进行创作，通过设计展现场地悠久的历史与文化，成为崇尚自然的领头军，成为亲近宇宙万物的推动力。

校园峡谷——梨花女子大学

Campus Valley—Ewha Womans University

翻译　王玲

该项目要求与整个校园以及 Shinchon 区的南部具有紧密的联系，这就需要创造出一种"设计效果大于实际面积"的视觉效应、城市效应和全方位的绿化效应，使其将梨花女子大学校园融入到首尔这座大都市中，这也是该项目的设计难点。原场地是校园的公共集会广场，如今新建的校园建筑、屋顶花园和一流的体育设施将其取代之，不仅满足了各种功能需要，同时也增强了它的社区归属感和兼容性。

该项目既别出心裁，又颇具争议。"校园峡谷"与狭长的体育场交相辉映，在该区域形成一个新地形，这种地形在某些方面又很大程度地影响着周围的景观。狭长的体育场和"校园峡谷"一样具有多重功能——既是通往梨花女子大学校园的新入口，又是人们举行日常体育活动和节日庆典活动的场所；同时，它还是真正连接校园与城市的元素。市民可以将这里作为休闲放松的场所，而学生则将屋顶花园和附近的体育场作为学习之外的乐园。最为重要的是，该项目不仅功能丰富，而且永远都充满着生机与活力。

狭长的体育场仿佛一条横幅，向 Shinchon 区的居民展示出丰富多彩的大学生活。体育场上行人穿梭、热闹非凡。新建大道的设计灵感来自香榭丽舍大街，它向下逐渐延伸至教学楼中心区，不仅引领学生和游客向北穿过校园中心，还将场地上的不同区域连接起来。这条步行大道的地势低于屋顶花园几层楼的高度。城市景观小品不时点缀其间，营造出别样的休闲空间。

连接原来建筑的道路被保留了下来，并且在"峡谷"之上新建了"桥梁"，从而打通了之前被梨花体育馆所阻隔的东西通道。一些新建的地下通道将校园中心与克莱拉教学楼、科思教学楼、菲佛教学楼、托马斯教学楼、吉比恩教学楼和健身房有机地连接在一起。这样形成了一个三维的连接体系，真正地将校园中心融入到整座校园以及更广阔的城市中。

田园风格或许是该校园最显著的特征——这里草木繁茂，百花争奇斗艳。屋顶花园的景观与环境优势彰显得淋漓

尽致。成为了人们休闲聚会及非正式课堂的绝佳场所。渗透交融的校园设计理念再一次得到了体现，从而模糊了新与旧、建筑与景观、现在与过去之间的界限。

"校园峡谷"展现出梨花女子大学校园中心的内部生活。如香榭丽舍大道般的公共空间仿佛一幅生动的画卷，将各种校园生活跃然纸上。它是一条限制交通流量的下沉式大道，人们行至大道的尽头又会拾级而上，走出地面。这与巴黎的香榭丽舍大道或者罗马的坎皮多里奥广场在美学设计上都有异曲同工之处。该项目是通向不同空间的门户，是通向其他地方的轨道节点，也是学生们课后探讨交流的场所，更是一个配有餐厅的休闲聚会的理想之所。同时还是一个充分利用了"峡谷"一端大阶梯的户外剧场。雕塑花园的规划也使这里充满了艺术气息，室内艺术活动轻松地延伸至户外，使其更具参与性与融合性。

正是这种恰到好处的灵活性（无论是概念上还是现实中）使得梨花女子大学校园中心能够很好地融入到景观中。它时而是建筑，时而是景观，有时候甚至可以被幻化成一尊完美的城市雕塑。

新校区的设计新颖独特，并力图成为为韩国培养未来女性领导人的摇篮，这在很多方面与美国的卫尔斯利女子学院极为相似。该项目鼓励大学及其学生追求卓越、专业和对周围事物的体察。

The complexity of the immediate site through its relationship to the greater campus and the city of Shinchon to the south demands a "larger than the site" response, an urban response, a global landscaped solution that weaves together the tissue of the Ewha campus with that of this city in metropolitan Seoul. The new university building, roof garden and state-of-the art sporting facilities occupy the site of what was earlier a communal meeting point, Ewha Square, justifying the need to build while maintaining a sense of belonging and inclusion in the community.

Dominique Perrault and Baum's design for the campus is simultaneously engaging and provocative. The campus valley, in combination with the sports strip creates a new topography for this area of the city that blatantly impacts the surrounding landscape in a number of ways. The sports strip, like the valley, is many things at once: it is a new gateway to the Ewha Womans University's campus, a place for daily sports activities, a ground for special yearly festivals and celebrations, as well as an area which truly brings together the university and the city. People in the city will use it as a leisure spot, gateway, and reference, while students will use the roof garden and adjacent sports strip as their place outside the academic bubble. The site is, most importantly, a place for all, animated all year long.

Like a horizontal billboard, the sports strip presents the life of the university to the inhabitants of Shinchon, and vice-versa. Once through the sports strip, pedestrian movement and flow through the site is celebrated. A new "Champs Elysées"-inspired boulevard that descends into the core of the academic buildings invites the public into the site carrying students and visitors alike through the campus' centre northwards, and bringing together the different levels of the site. This pedestrian highway is located on a lower level, several storeys below the roof gardens. However, it is once again a site for conviviality, with urban furniture provided at intervals.

The paths connecting the existing buildings are maintained, with new "bridges" crossing the valley and creating new east west connections that were previously limited by the Ewha stadium. Underground connections are also suggested, connecting the campus centre to Clara Hall, Case Hall, Pfeiffer Hall, Thomas Hall, Gibeon Hall, and the Gymnasium. A three-dimensional system of connections is therefore possible, truly integrating the campus centre with the campus, and once again with the broader city.

The pastoral nature of the campus is perhaps its most remarkable quality. It should be permitted to grow outwards, or inwards in this case, covering the campus centre with trees, flowers, and grass. The landscaping and environmental advantages of roof gardens are all there. The park is re-drawn, resulting in an idyllic garden and creating a special place for gathering, conducting informal classes, and simply relaxing. The notion of weaving together the campus is again evident, blurring the distinction between old and new, building and landscape, present and past.

A new seam slices through the topography revealing the interior of the Ewha campus centre. This "Champs Elysées"-style public space shapes a void is formed and constructs a hybrid place in which a variety of activities can unfold. It is, simultaneously: an avenue, gently descending, controlling the flow of traffic, leading to a monumental stair carrying visitors upwards, thus establishing a visual/aesthetic connection to the Champs Elysées on Paris or the Campidiglio in Rome; it is also an entry court, from which access

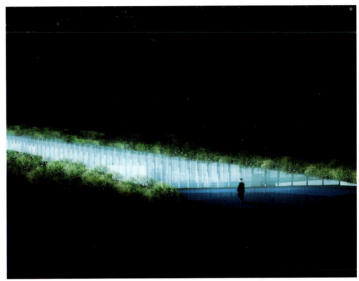

to the various departments exist; a node, or point on a trajectory to another destination; a forum for the exchange of ideas as students gather after class to discuss their views; a piazza, with the cafeteria spilling out creating a real "place" to stop and relax. It is also an outdoor theatre or amphitheatre, making use of the grand staircase on one of the ends of the canyon. Finally, a connection to art is achieved by including a sculpture garden program, where indoor gallery events can very easily be pushed outwards, and made more engaging and inclusive.

It is precisely this flexibility (conceptual and real) that allows the new Ewha campus centre to inevitably weave itself into the landscape, as a building, a landscape at other times, and every so often a pure urban sculpture.

The new campus clearly sets Ewha as a ground-breaking, forefront university for future female Korean leaders – in many ways similar to Welesley University in the United States. The sense of direction given to the institution and its students by this project is one of excellence, expertise, and awareness to one's environs.

项目位置：韩国首尔市
客　　户：梨花女子大学
项目预算：1 亿美圆
占地面积：19 000 ㎡
建筑面积：70 000 ㎡
建筑体积：350 000 ㎥
景观面积：31 000 ㎡
建设时间：2004 年 ~2008 年
建筑设计 (团队主管)：Dominique Perrault 建筑设计事务所 (巴黎)
当地建筑合伙人：Baum 建筑设计事务所 (首尔)
景观设计：CnK Associates (首尔)

Location: Seoul, South Korea

Client: Ewha Womans University

Budget: US$100 million

Surface Area: 19,000 ㎡

Built Surface: 70,000 ㎡

Built Volume: 350,000 ㎥

Landscaped Area: 31,000 ㎡

Project Dates: 2004 ~ 2008

Architecture (Team Leader): DPA Dominique Perrault Architecture (Paris)

Local Architecture Partner: Baum Architects (Seoul)

Landscape Design: CnK Associates (Seoul)

绿地走廊——东京中城景观设计

Greenbelt—Landscape Design for Tokyo Midtown

翻译　王玲

　　2002 年 3 月～ 2003 年 8 月间，考古工作者对这里展开了考古分析，并发现了超过 50 000 件江户时代（1596 年～ 1698 年）的陶器和两枚金币。

　　东京中城不仅是迄今为止东京最大的多功能项目之一，还是万众瞩目的东京新地标。它涵盖了办公、住宅、商业、酒店和其他高品质城市空间，面积达 557 418m²，这些空间与地下公共交通系统有效地接驳。248m 高的中心建筑——东京中城大厦，是日本第二高建筑。整个项目占地面积为 101 000m²，其中包括对拥有 400 年历史的桧町公园进行改造。

　　该项目的设计目标是希望在景观和建筑设计过程中以一种现代方式捕捉〝日本〞的本质。因此，客户希望创造出一个极具表现力的景观设计——既不是建筑的衍生物，也不要太深地扎根于日本文化。

　　该项目的景观设计标志着日本景观发展的一个巨大转变。项目中有超过 50% 的场地都是开放空间或者绿地，这对于甚至连步行道或者可以让人们坐下来休息放松的场所都鲜有的东京显得弥足珍贵。丰富的景观元素与周边的开阔空间和绿地交相辉映，演绎出建筑内外自然流畅的空间。新景观设计的最大亮点之一就是将场地上原有的 140 株成熟的树木保留并重新种植。

　　设计师提出的〝绿地走廊〞勾起了人们对场地历史的记忆。一条曾经横贯场地的历史上有名的河流也被重新演绎——设计成浪漫的现代水景。人们被吸引到公园，沿着蜿

蜓的小路前行至 21_21 设计博物馆、大草坪和传统日式庭院。行人天桥在满足公园空间的需求之外，还在设计中增强了渗透和联系。建筑景观与抽象的自然设计时而并列、时而交错，给人们带来一场充满活力的视觉盛宴——广场上弯曲的建筑天窗和公园里河流上覆盖的线条都是抽象的自然设计形式。

多层设计理念也被运用到该项目的设计中，形成别具一格的空间序列。设计出发点是基于并强调场地的原始特征，如场地的地形变化和场地上葱郁的树木。这些场地的原始特征都被重新规划，并与被重新诠释的场地历史文化和自然历史的新元素相结合，共同为人们营造出丰富的身心体验和独特的地域感。

东京中城的大草坪在日本是一个不寻常的景观设计，因为日本的设计标准通常是小地块的设计。大草坪是寓建筑表达于景观之中，它如行云流水般巧妙柔和地融入公园之中。设计师打造了一处开阔的绿色空间，这里不仅可以举办各种活动，更有益于人们修身养性或培养园艺情趣。40 株成熟的樱花树被保留并在场地上重新栽种，它们在与桧町公园入口通道连接的同时形成了一条赏心悦目的"樱花长廊"。在日本，樱花盛开的时节也正是聚会频繁的时刻——人们远离工作，带上毯子、野餐用具和各种饮料相聚在樱花绽放的树下。每到这个时候，成千上万的游客便会纷至沓来，享四季之更迭、赏樱花之美景。

比例尺 1:5000

Archaeological exploration (once a feudal estate) was conducted from March 2002 to August 2003, during which period over 50,000 pieces of Edo-period (1596~1698) pottery were found, along with two gold coins.

The 25-acre Tokyo Midtown Project is one of the largest mixed-use projects undertaken in Tokyo to date, and a world-class mixed use complex that has become a landmark for the City of Tokyo and the premiere address in Japan. The new development includes 6 million square feet of office, residential, retail, hotel and high-quality urban space – linked to mass transit below. The central building, Tokyo Midtown Tower, is the second tallest building in Japan at 248 meters (814 feet). The total development totals 10.1 hectares (25 acres), including a refurbished, 400-year-old site (Hinokicho Park).

The design goal was to capture "Wa", which translates as the essence of "Japanese-ness", in a modern way in the landscape and the architecture of the development. The client wanted a

strong, expressive landscape that was neither derivative of the architecture nor too deeply rooted in Japanese culture.

The landscape design of the Tokyo Midtown Project marks a significant shift in Japanese development, as over 50 percent of the site's land stands as open or green space, precious assets in a city where sidewalks and places to sit and relax are few and far between. The complex's diverse elements were designed to be in dialogue with the surrounding open-space and expansive greenery, creating a flow between the outdoor environment and the interior of the complex. One of the most notable features of the new landscape design is the 140 mature trees that were preserved and transplanted on the site.

EDAW's design for "the greenbelt walkway" recalls the history of the site and a historic stream that once traversed the site is reinterpreted as a contemporary, romanticized water feature, drawing people into the park and leading down though meandering pathways to the 21_21 Design Museum, the Great Lawn, and the traditional Japanese Garden. Pedestrian bridges jut out to meet

the park spaces, reinforcing the notion of interpenetration and connectedness. The juxtaposition and weaving of architectural landscape and abstracted natural forms – the curving architecture of the skylights in the plaza and straight lines overlaid onto the meandering stream in the park – creates visual dynamism between elements.

Multiple and layered design ideas have blended at Tokyo Midtown to create a unique sequence of open spaces. The design's starting point involved building upon and emphasizing existing site characteristics, such as topography and a lush, mature tree canopy. These elements were reconfigured and layered with new elements that reinterpret the site's cultural and natural histories, carefully focused to create rich experiential qualities and a distinct sense of place.

Tokyo Midtown's Great Lawn is a particularly unusual feature in Japan; the standard for Japanese design is usually on a more intimate scale. It is an architectural expression in the landscape, overlaid onto the softer, stream-like forms of the park. The designers wanted to provide open, inviting green spaces that can be programmed for events as well as ones more conducive to solitude and horticultural interest. 40 mature cherry trees were preserved and transplanted on site, linking the entryway to Hinokicho Park and creating a "cherry promenade." Cherry blossom season is a storied time of celebration in Japan. When the trees are in bloom, people take off time from work to gather at their favorite tree with a blanket, picnic, and drinks. During this season, Tokyo Midtown can expect thousands of visitors attracted by the opportunity to celebrate the cycle of seasons and the beauty of the trees.

項目位置：日本东京六本木区

客　　户：三井不动产株式会社

预　　算：31 亿美圆

项目时间：2002 年（设计）；2004 年－ 2007 年

总 面 积：563 800m²（博物馆／特殊区域：20 300m²）

景观设计：易道建筑景观设计事务所（美国旧金山）

建筑设计：SOM 建筑设计事务所（美国纽约）

　　　　　Communication Arts 公司（美国科罗拉多州）

　　　　　Fisher Marantz Stone 照明设计事务所（美国纽约）

　　　　　Buro Happold 公司（美国纽约）

　　　　　株式会社日建设计（日本东京）

　　　　　隈研吾建筑设计事务所（日本东京）

　　　　　坂仓建筑工程研究所（日本东京）

　　　　　青木淳建筑规划事务所（日本东京）

　　　　　安藤忠雄建筑设计事务所（日本大阪）

所获奖项：2008 年城市土地学会全球奖

　　　　　2007 年国际地产投资交易会亚洲类别最佳：

　　　　　不动产创新和设计奖

　　　　　国际地产投资交易会亚洲最佳多功能开发奖

Location: Roppongi District, Tokyo, Japan

Client: Mitsui Fudosan Co., Ltd.

Budget: US$3.1billion

Project Dates: 2002 (design); 2004~2007

Surface Area: 563,800 m²(Museums/special areas: 20,300 m²)

Landscape Architect: EDAW Inc (San Francisco, USA)

Architects: Skidmore, Owings and Merrill (New York, USA)

　　　　　Communication Arts, Inc. (Colorado, USA)

　　　　　Fisher Marantz Stone (New York, USA)

　　　　　Buro Happold (New York, USA)

　　　　　Nikken Sekkei, Ltd. (Tokyo, Japan)

　　　　　Kengo Kuma & Associates (Tokyo, Japan)

　　　　　Sakakura Associates Architects and Engineers (Tokyo, Japan)

　　　　　Jun Aoki and Associates (Tokyo, Japan)

　　　　　Tadao Ando Architect and Associates (Osaka, Japan)

Awards: 2008 Winner, Urban Land Institute (ULI) Global Award

　　　　2007 Winner Overall Best in Class: Property Initiative and Design,

　　　　MIPIM Asia Award

　　　　Best Mixed-Use Development, MIPIM Asia Award

卡塔尔教育城景观设计

Landscape Design for Qatar Education City

翻译　王玲

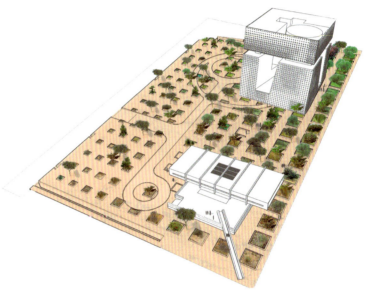

教育城的图书馆大楼仿佛天外来客般在其四周留下许多不均衡的线性空间。当人们在图书馆附近看到复杂的城市规划、热火朝天的建筑开发以及在建筑和花园中所使用的丰富多样的不规则造型时，不禁会深刻体会到简化新公园设计的必要性，而非一味地追求复杂化。

设计师提出了一个简明的花园设计方案——一组组圆形种植槽印刻在地面土壤上，里面种植着三种不同的耐干旱植被：阿拉伯树胶、大戟属植物和龙舌兰属植物，凹陷的圆形种植槽则是按照植物大小和深度变化进行排列。如果同类植被较少的话，就将它们种植在较浅、较小的种植槽里；较多的话，则种植在较深、较大的种植槽里。这样不仅给人们带来了阵阵惊喜，而且还尽量利用同种植被来营造出变换的景观效果。

设计的目标之一是仅在头几年里对肉质植物和树木等植被进行浇灌，当这些植被成活下来后就会停止人工浇灌。花园也充当一个教学示范区，可以在当地气候和土壤条件下对各类植物进行研究。圆形种植槽之间的区域是花园主要的铺装带，将当地石灰岩切割成粗犷的矩形块嵌入沙中。尽管这种沙色的铺装显得有些凌乱，却非常适宜人们行走；圆形种植槽则生动地点缀其间。

主楼花园中汇集了各种植被，包括本地树木、灌木、开花植物和块茎植物等耐热、耐蚀、耐干旱植被。这些植被不仅呼应了一条精心设计的南北向分级供水系统，还营造出了变换多端的花园景观；无论是苍翠葱茏还是落叶疏枝，无不别具风情。

设计中还战略性地融合了许多来自与卡塔尔极端气候相似的国家的不同植被，它们有效地丰富了景观形式，增强了遮阴、气味和颜色效果。主楼花园位于相同尺寸停车场之上，由钢筋和石材铺就而成，这种特定的"屋顶花园"最多只能承受1.2m深的土壤，因此无法为植被提供足够的土壤层。为了实现花园的设计目标，设计师在平坦坚硬的"屋顶"上有规则的放置一些方形植栽容器，这些植栽容器自南向北依次变换着大小、高度和间距。

方形植栽容器大小不一，其边长从6.6m～14.4m不等，深度从0.45m～1.2m不等。它们形成丰富多样的植物群落，每一个植物群落又由各种相互影响、相互促进的不同植被所组成。植物在精心设计的土壤层中茁壮成长，并根据需要增设了挡水和养分补充设计。此外，这些方形植栽容器形成一个装饰面，仿佛色彩斑斓的地毯，将地面上所有的功能服务设施融为一体。这里不仅是拔地而起的建筑的"明信片"，还是吸引游客纷至沓来的美丽花园——无论是在楼内或街道上观赏到的日间景致，还是夜晚的花园景致。

Lush flowering trees

Sculptural trees+grasses

Grasses and shrubs

Grasses, shrubs and perennials

Desert perennials

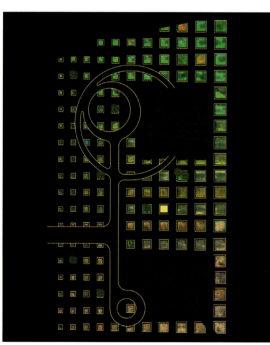

ENDEMIC PLANTS, SHRUBS, TREES, GRASS, BULBS & SUCCULENTS
IN COMPOSED GROUPINGS ARE WATERED TO DIFFERENT DEGREE
IN RICH TO ARID SOIL TO STUDY PLANT BEHAVIOUR

100   300   500

The library building seems to have landed from elsewhere onto the site, leaving uneven amounts of linear space around its four sides. Looking at the city plan in the vicinity of the Library and recognizing the amount of development and the diversity of irregular shapes used for buildings and gardens, one tends to feel the need to simplify rather then complicate the design for a new garden.

Inside Outside therefore proposes a simple garden recipe: a group of circular imprints into the soil, planted with families of three different types of drought tolerant vegetation:

Acacia, Euphorbia and Agave. The effect of the logic we use for the concave, circular imprints – their changing size and depth – will be, that less of the same species will grow on a higher view level in the smaller areas; and more of the same species will grow on a lower level out of the larger circular areas…thus providing surprise and change with a minimum of variation.

Our aim is to water the plants, succulents and trees only in the first few years – until they have settled – and to refrain from watering the plantings after this period. Thus the garden will become an educational one, where various species of one family can be studied in the local climate and soil conditions. The main surface of the garden – the area in-between the imprints – is made of local limestone, cut into rough rectangular blocks and embedded in sand. This will create a slightly irregular but walkable plane in the color of sand, in-between round concave planting areas.

The idea for the headquarters' garden is to consider it as a

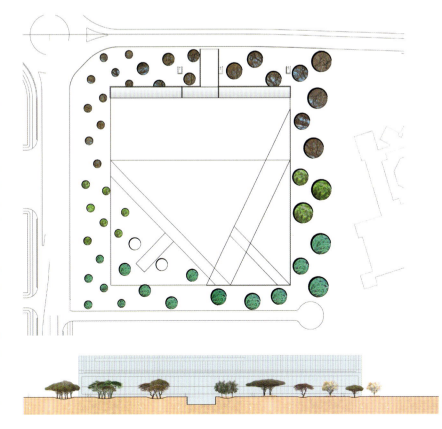

collection of habitats, planted with indigenous trees, shrubs, flowering plants and bulbs - tolerant of heat, salty wind and draught – that will react to a carefully measured gradational water system that is organized in north-south direction; thus providing a rich variety of gardens, from green and lush to sculptural and dry. Species from other countries of the same hardiness zone as Qatar will be mixed in strategically, to add form, shadow, scents and color effects. As the future garden is, in fact, a floating stone and steel plate above a parking lot of the same size, it does not offer full soil to its planting. The given "roof top" situation allows for a maximum of 1,200 mm soil depth only. To create the garden that we envision, we lay down a regular system of square planters over the flat, hard roof surface, that gradually change in size, height and distance in north to south direction.

The generous sizes of the planting beds – from 14.4m to 6.6m square by 0.45m to 1.2m depth – give the opportunity for a broad series of variegated biotopes, each built up of a variation of plants, shrubs and trees that complement and stimulate each other; and that grow in carefully composed build-up of soil and – if so required - water retaining elements and added nutrients. At the same time, this system of planted squares will form one decorative plane, a colorful carpet that unites all the functional services of the ground plane – while offering an inviting platform for the buildings that rise from it and an attractive garden for visitors; both during the day (seen from many positions inside the buildings and from the street) and at night (when used as garden).

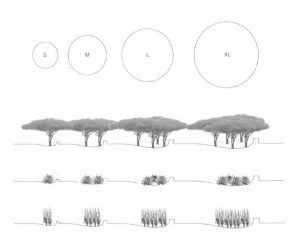

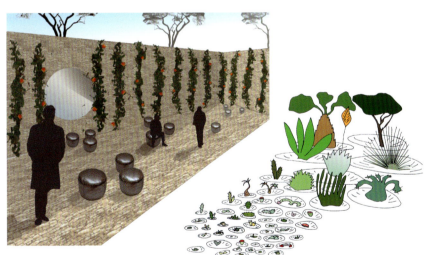

项目位置：卡塔尔多哈

面　　积：43 000m² （图书馆）　38 000m²（主楼）

项目时间：2008 年～在建

客　　户：卡塔尔教育科学与社会发展基金会

建筑设计：荷兰大都会建筑事务所（鹿特丹）

景观设计：Petra Blaisse（Inside Outside 景观设计事务所）

Location: Doha, Qatar

Site Size: 43,000sqm (library), 38,000sqm (headquarters)

Project Dates: 2008~ ongoing

Client: Qatar Foundation for Education, Science and Community Development

Architect: OMA (Rotterdam)

Landscape Design: Petra Blaisse (Inside Outside)

低碳景观——法贝公园

Low-carbon Design—Faber Park

翻译　李沐菲

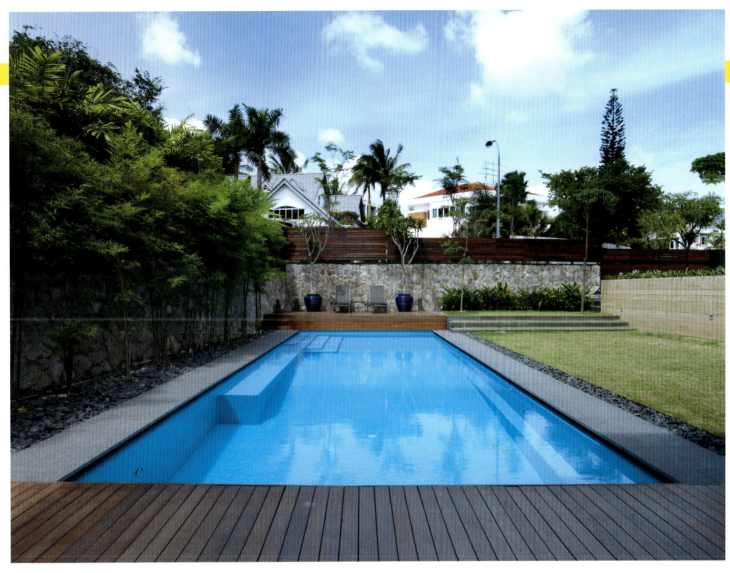

该项目最迷人之处就是深灰色的铝锌合金饰面，远看就如同一个黑盒子。在有着无数类似房屋的街区里，该项目无疑是一个独特的建筑表述。从外表看其具有简洁的外形，而内部则融入了客户的意见——洋溢着温馨、惬意以及现代的气息。

该项目建筑面积达 591.55m²，可供一个五口之家居住。主要的设计理念就是希望创造巨大的开放式空间以及充满绿意的热带环境。客户要求该项目是一个环保型住宅，能够拥有充足的光照，并且通过各个空间的相互连接实现良好的通风。因此，设计师采用了白色的水磨石地面，使其看起来干净、轻盈。为了进一步实现空间开放这一目标，建筑内部的许多区域都能够不着痕迹地向相邻的空间敞开，如卧室可以开放并与餐厅和游泳池相连，这样可以使客人在游泳池旁边享用晚餐；厨房的入口处安装了隐形滑道门，并且能够推进壁橱后面，使厨房与餐厅瞬间变为一个整体空间。

建筑内部所设置的一座巨大的殖民地风格的旋梯成为了

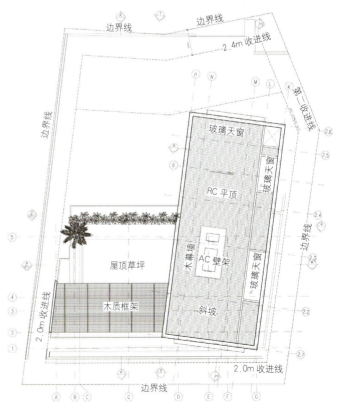

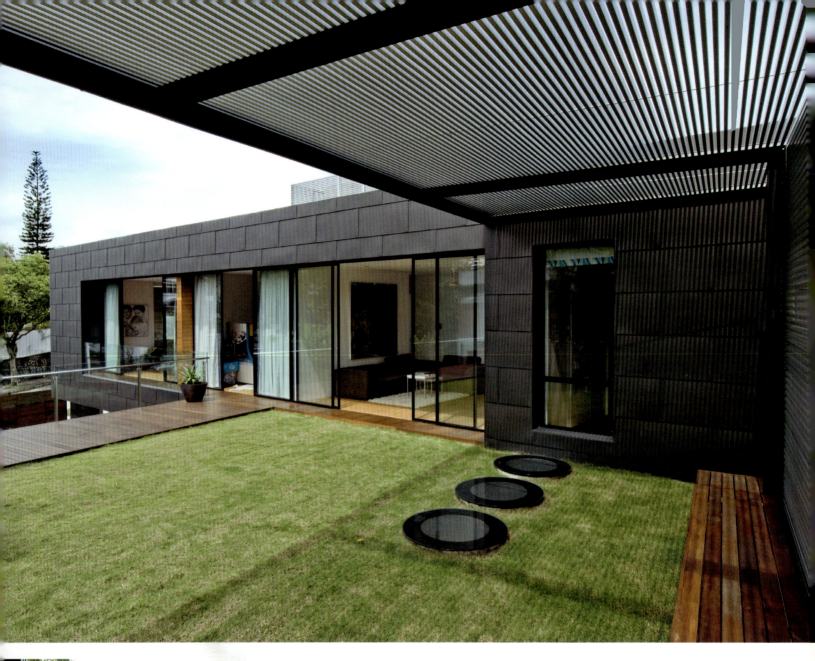

该项目的视觉焦点，旋梯通往住宅中的公共区域，其中包含一个主卧室、一个带有两间卧室的套房以及一个单独的普通卧室。温馨宜人的卧室对于住宅显得尤为重要，因此二楼整体使用了柚木地板。为了加强开放空间的叠加效果，家庭娱乐室正对着带有烧烤设备的花园，并且可以俯瞰园内的泳池，四周竖起的铝质屏风为住宅提供了必要的私密性，为了呼应其环保的主题，在浴室的顶棚增设了天窗，以减少对照明能源的需求。浴室内的装修则采用了黑色的花岗岩、银色的石灰华和白色的大理石。女用卫生间的轮廓一半由墙体构成，另一半则点缀生机盎然的绿色植物，并装有金属的卫浴饰品，具有浓浓的"花园式"风格。

地下室的设计为该项目提供了更多的娱乐空间，其中包括一个家庭影院、一个儿童活动室以及一个设有台球桌和吧台的空间。其内部空间采用了木质风格的装饰，且板材及饰面都设计得十分灵活。为了避免地下室光线不足以及通风不好的问题，设计师在泳池和地下室之间设计了一个庭院——地下室的隔墙采用玻璃材质，这样便可以看到泳池，使人如同置身水族馆一样。总之，从各个角度看，该项目的室内与室外空间相互贯通，不分彼此。

The most attention-grabbing detail of this project is the dark grey metal cladding made of aluminum and zinc resembling a black box from afar. It is meant to be an architectural statement amid a myriad of similar houses in the neighbourhood. It oozes edginess from the facade yet contains a warm, cozy and modern interior to accommodate the client's lifestyle.

The Gross Floor Area (GFA) of this magnificent-looking house is 591.55m^2 and resides a family of five. The concept brief was to have huge open spaces with lots of greenery in a tropical environment. The client also requested for an eco-friendly house that lets in plenty of sunlight and has to be well-ventilated throughout the inter-connected spaces. The result was white terrazzo flooring that gave it a cleaner, airier look. To further implement the concept of open space, many of the areas can be opened up into the adjacent space seamlessly. For instance, the living room opens up into the dining room and swimming pool. This allows the client to be able to enjoy their dinner by the poolside. The entrance of the kitchen was fitted with sliding doors that were concealed in hidden pockets and can be pushed back to allow seamless integration into the dining area.

Inside the immaculate interior, a huge, spiral staircase is constructed defining an air of colonial style. This forms the focal point in the house. The staircase opens out to a family common area and consists of a master bedroom, one junior suite with two bedrooms and a separate common bedroom. Having a warm and inviting bedroom space was very important. Therefore, teak plank flooring was used throughout the second level. To reinforce the concept of open spaces overlapping, here, the family room opens out to a garden with a barbeque pit and overlooks the pool. Aluminum screens were erected to provide privacy from the neighbouring homes. In line with the eco-theme, a skylight was fitted in the bathroom to reduce the use of energy for lighting. The play of black granite, silver travertine and white marble was used in the bathroom. The powder room is adorned with silver and metal accents and has a "Garden" feel to it because it is only partially covered by walls and the rest of the space is decorated with greenery.

A basement was conceived to allow more space for recreation. This basement houses a home theater, kids' playroom, a corner for the billiard table as well as a bar. There is a consistent accent

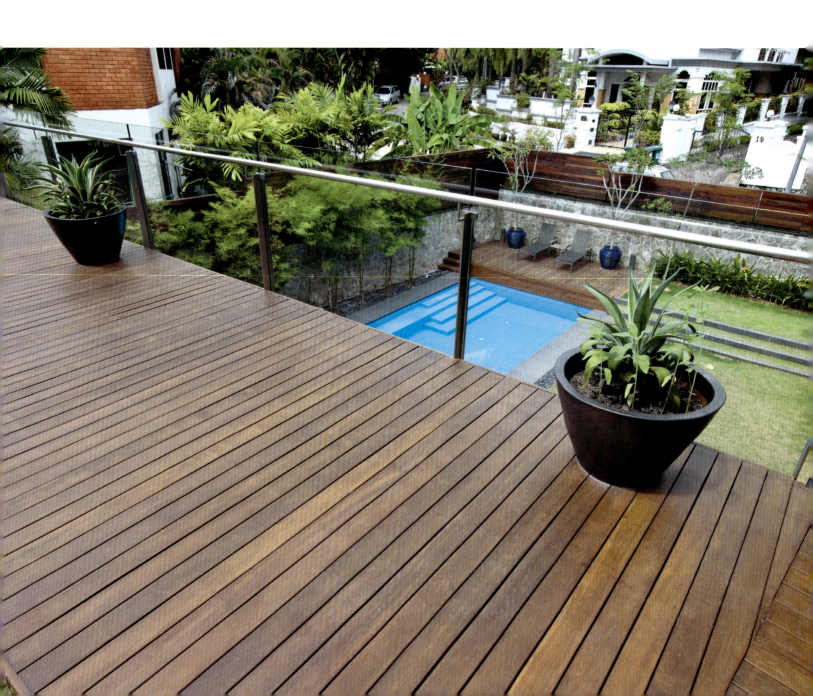

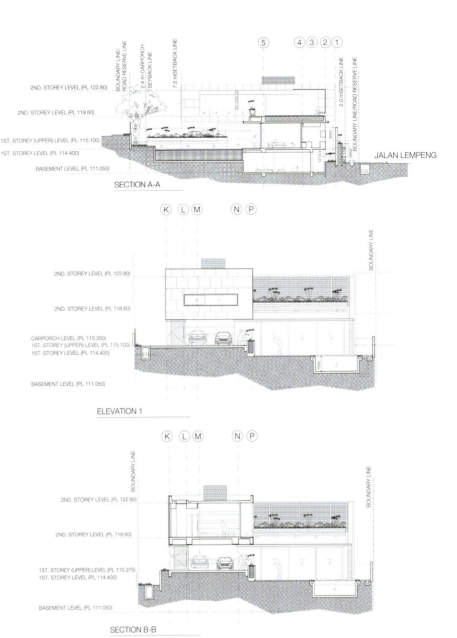

2ND. STOREY LEVEL (PL 122.80)
2ND. STOREY LEVEL (PL 118.60)
1ST. STOREY (UPPER) LEVEL (PL 115.100)
1ST. STOREY LEVEL (PL 114.400)
BASEMENT LEVEL (PL 111.050)
JALAN LEMPENG

SECTION A-A

2ND. STOREY LEVEL (PL 122.80)
2ND. STOREY LEVEL (PL 118.60)
CARPORCH LEVEL (PL 115.350)
1ST. STOREY (UPPER) LEVEL (PL 115.100)
1ST. STOREY LEVEL (PL 114.400)
BASEMENT LEVEL (PL 111.050)

ELEVATION 1

2ND. STOREY LEVEL (PL 122.80)
2ND. STOREY LEVEL (PL 118.60)
1ST. STOREY (UPPER) LEVEL (PL 115.375)
1ST. STOREY LEVEL (PL 114.400)
BASEMENT LEVEL (PL 111.050)

SECTION B-B

of veneer and solid timber used liberally in the basement. In order to avoid having the same problem that most basement areas face – insufficient sunlight and ventilation – a courtyard was created between the swimming pool and the basement. Sunlight streams in from the courtyard and keeps the area well-ventilated. From the basement, a glass wall is erected, overlooking the pool, conceiving notions of being in an underwater aquarium. Throughout the house, from almost every angle, the inside is outdoors almost as much as the outside is indoors.

项目位置：新加坡
客　　户：Ms Sue Catton
建筑面积：591.55 m²
景观设计：Ong & Ong
结构工程：KKC 咨询服务公司
建成时间：2008 年

Location: Singapore
Client: Ms Sue Catton
Surface Area: 591.55 sqm
Landscape Design: Ong & Ong
Structural Engineering: KKC Consultancy Services
Completed Time: 2008

私人沙漠度假胜地
Private Desert Retreat

翻译　王玲

该项目所面临的挑战是要在恶劣的沙漠气候条件下，创造一个能够舒适地举行各种大型活动的户外场所。此外，为了满足客户夏季度假的需求，该项目的设计及建造必须在6个月内完成。

该项目位于叙利亚快速发展的 Yaafur 地区，占地面积为15 000m²。项目的场地干燥平坦，包括一座原来的房子、一个旧游泳池和一座员工住房。客户来自一个颇具声望的叙利亚家族，他希望建造一座私人度假胜地，将干燥荒凉的场地改造成一片沙漠绿洲。项目的景观设计独具特色，富于视觉冲击力，优先采用了舒适、耐用和高品质的材料。

客户对于项目的预算没有特别限制，因此，设计师着手打造一个能够适应沙漠极端气候的独特场所——包括改造花园、替换原来的游泳池，以及建造一个可以容纳200人的舒适场所。

两面砂岩墙与房子垂直设置，它们不仅可以将员工住房隐藏起来，同时也将项目引入新的花园和景观带。一个安装在墙面的自给式定制设备最大限度地满足了项目的设计挑战。所有的功能空间，如厨房、露天酒吧、卫生间、淋浴间和更衣室都被一系列线性空间的墙壁所包围。在墙壁间设有一处水景，其在视觉上将墙壁连接在一起，不仅为游客营造出宾至如归的感觉，而且潺潺的水声仿佛悠扬的小夜曲，令人心旷神怡。新泳池和花园所营造出的丰富景观为悬臂式绿廊带来了丝丝阴凉与呵护。

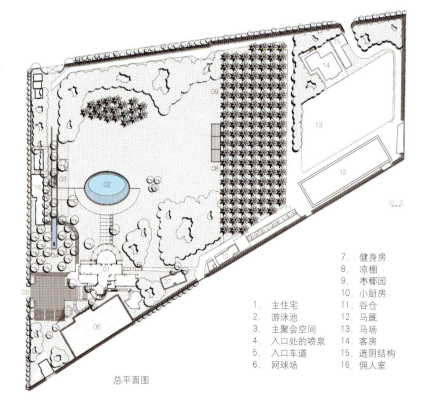

1. 主住宅
2. 游泳池
3. 主聚会空间
4. 入口处的喷泉
5. 入口车道
6. 网球场
7. 健身房
8. 凉棚
9. 枣椰园
10. 小厨房
11. 谷仓
12. 马厩
13. 马场
14. 客房
15. 遮阴结构
16. 佣人室

总平面图

为了进一步减弱恶劣的沙漠环境所带来的影响，遮阳屏和气候控制系统被分散地安装在建筑结构中。遮阳屏沿着线形沟槽排列，热空气和冷空气通过其他沟槽释放出来。隐藏式的音乐和照明系统进一步增强了这里的宁静与祥和。

新泳池位于原住房的轴线上，蓄满水的石灰华泳池成为了开展聚会和其他活动的背景幕。在纯正的椭圆形泳池中，巨大的石灰华板整齐地排列在一起，并由当地的石匠手工雕刻成无缝的优雅形式。这种形式使泳池微微倾斜，形成一个水吧，继而形成一个水下悬浮休息区和按摩池。

该项目仅采用了3种材料：铺装路面和泳池的石灰华、绿廊的砂岩墙面以及遮阳结构的柚木板。

照明设计推陈出新，完全改变了传统的夜间氛围，打造出令人流连忘返的夜景。完全可控的照明系统随着夜晚的推移可营造出不同的场景，无论是朦胧的灯光还是为集会而设计的动感四射的灯光，都可以通过一系列预设装置来实现。装有8mm端点纤维和透镜的光纤灯被嵌入到泳池石灰华之间，使整个水面产生了微妙的色彩变换。

嵌入式向上照射的光纤灯和线形嵌入式向下照射的光纤灯是该项目的主要照明策略。灯饰被完全隐藏起来，这是整个设计不可分割的一部分。

设计专门选用了一些适合干燥气候的植被。而园景树的组团种植同时起到了提升项目和围合空间的作用。位于园景树后方的枣园提供了超过15种的食用枣品种。枣树荫蔽着一个巨大的玫瑰园，开花时节花香四溢，采摘下的玫瑰花还可以装点聚会等活动。幸运的是场地上有一口天然水井，因此草坪——这种客户最中意的地面覆盖物——可以得到广泛应用。在夏日的拂晓和黄昏时分，可以通过自动喷头进行浇灌草坪。

如今，花园里枝繁叶茂，鸟语花香；墙壁间种植的橡树华盖如伞，绿树成荫。所有的室外家具都来自 Dedon 休闲系列，同时也丰富了该项目的表现形式。

自建成以来，该项目满足了各种聚会的需求，促使客户对原有别墅进行了大规模的改造。别墅的改造采用了相同的景观材料，使房屋和花园融为一体，并增加一个私人温泉疗养场所。

The challenge of this project was to create an outdoor environment capable of comfortably hosting numerous large-scale events in a harsh desert climate. In addition, the entire project had to be designed and built in 6 months to be ready for the summer season. The project is located in a rapidly growing area of Syria, Yaafur, close to the border with Lebanon. The site, a 15,000m² dry and flat desertscape contained an existing house, an old pool, and an unsightly staff housing structure. The client, from a renowned Syrian family, wanted to create his private getaway and convert this dry desolate area into a desert oasis, setting an exclusive, immediate impact landscape, with comfort, durability, and high quality materials as a priority. The budget was open and was not an issue.

Hence, we set out to create a unique gathering environment, catered to the extreme desert temperatures. Our project consisted of revamping the gardens, replacing his existing swimming pool, and creating an area capable of hosting up to 200 people in total comfort.

Two sandstone walls, anchored perpendicularly to the house, conceal the staff housing and orient the project to the new gardens and vistas. Emanating from these walls, a totally self-contained, tailor-made facility was conceived to meet this challenge in the most efficient way possible. All functions required, such as industrial kitchen, open bar, bathrooms, showers, and changing rooms, are embedded within the walls, between series of linear gardens. A water feature visually connects the walls while providing a welcoming gesture to the garden and serenading the users with its soothing surrounds. A cantilevered pergola provides shade and shelter with uninterrupted vistas of the new pool and gardens.

To further counteract this harsh environment, shade screens and a climate control system are discretely incorporated within the structure. Screens slide through linear grooves, while hot and cold air is delivered through others. Concealed music and lighting systems further enhance the soothing character of this escape.

In axis with the existing residence, the swimming pool was conceived. A travertine stone basin brimming with water becomes the magical backdrop of gatherings and events. In this pure elliptical form, massive travertine slabs were laid side by side, then hand-carved by local stone masons to produce this seamless and elegant form that gently slopes to produce an aqua lounge, then warps to become an underwater suspended sitting area and Jacuzzi.

The entire project was built in three materials only - travertine stone for all flooring and pool, sandstone for the pergola walls, and teak wood for the shade structure.

The lighting concept was conceived with intent to totally transform the ambiance at night and turn the project into a memorable nocturnal scene. Being totally controllable, the project unfolds into various scenes as the night progresses with a range of preset options to cater to the particular type of event - from very subtle intimate lighting, to dynamic lighting for major gatherings. Fiber optics, equipped with 8mm endpoint fibers and lenses, embedded between the travertine stone in the pool, because the entire body of water to glow into subtle transitions of color.

Fiber optics, recessed uplights, and linear recessed downlights are the main elements used in the lighting strategy. Lighting was totally concealed and incorporated as part of the design.

Plants suitable for this dry climate were used exclusively in the project. Specimen tree groupings are incorporated to enhance and frame the project, along with a date tree orchard in the background to provide more than fifteen varieties of edible dates. Underneath that orchard, a vast rose garden provides scent and cut flowers to animate the gatherings and events. Luckily, the site was blessed

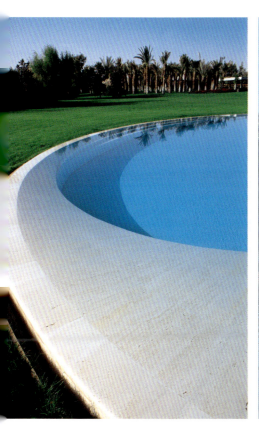

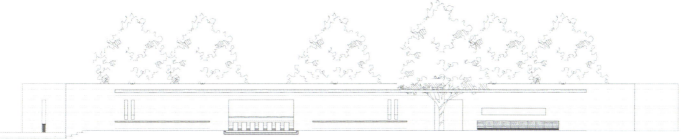

PERGOLA ELEVATION_SCALE 1/75

with a natural water well, hence, grass, the client's favorite ground cover, was used extensively and watered by automated sprinklers at early dusk and dawn in summer.

Today, four years after its completion, the gardens have fully flourished and the oak trees embedded between the walls have developed into impressive shading canopies. Entire outdoor furniture selection was from the Dedon lounge series, which complemented the project's vocabulary.

The project, since its inception, has been relentlessly catering for numerous family and friends gatherings, and has stimulated the client to currently undergo a major renovation on the existing villa utilizing the same landscape materials to render the house and garden as one entity, as well as adding a private spa extension.

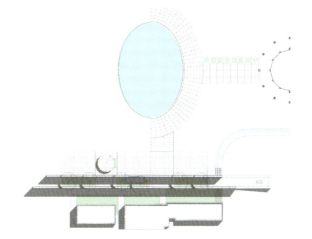

项目位置：叙利亚 Yaafur 地区
景观设计：Vladimir Djurovic 景观设计事务所
建成时间：2004 年 8 月
占地面积：3000m²
底层面积：35m²（绿阴露台）
总建筑面积：420m²（绿阴露台和地下室）

Location: Yaafur, Syria
Landscape Architect: Vladimir Djurovic Landscape Architecture
Completed Time: August 2004
Site Size: 3,000 m²
Ground Floor Area: 35 m²(shaded terrace)
Total Combined Floor Area: 420 m² (shaded terrace and basement)

Oceania 大洋洲篇

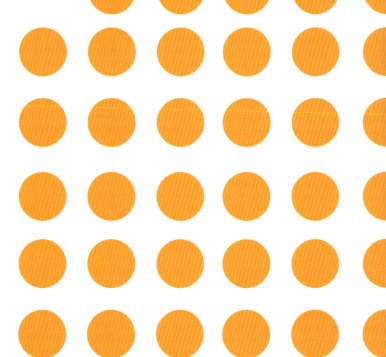

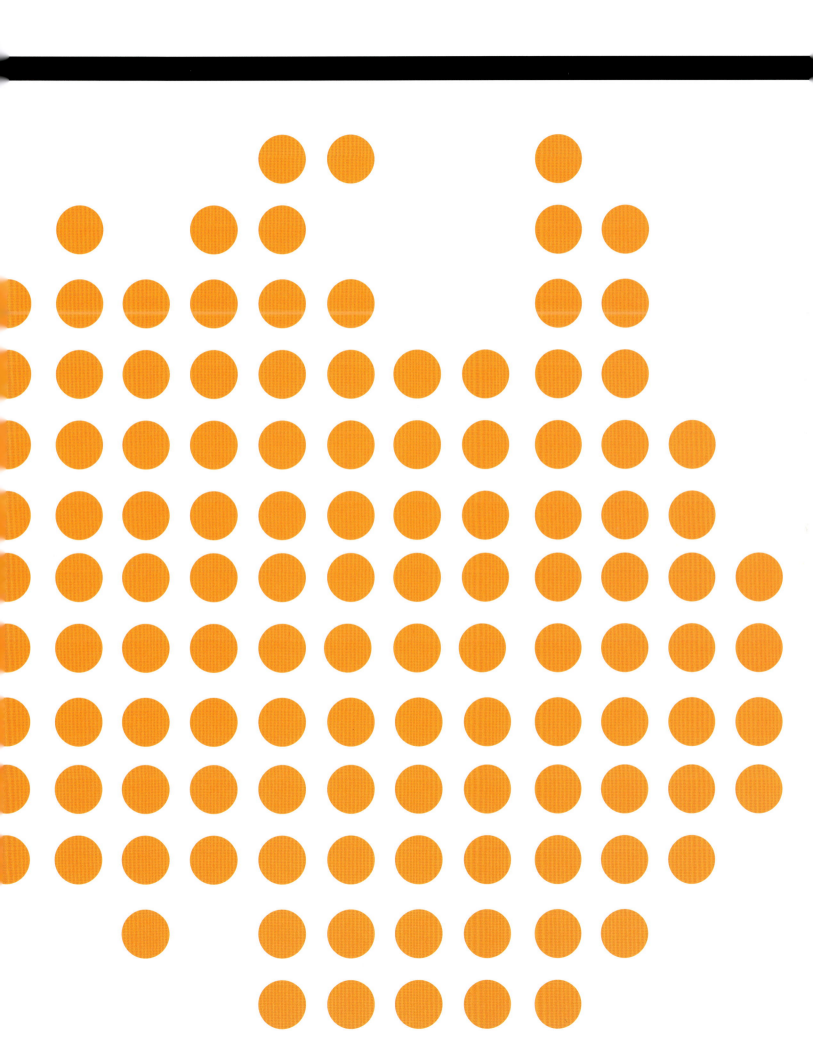

Perry Lethlean

Adjunct Professor for School of Architecture and Design, RMIT University
Principal, B.L.Arch MD (Urban Design) AAILA
Taylor Cullity Lethlean (Melbourne and Adelaide, Australia)
www.tcl.net.au

Successful projects demonstrate the ability to conceive of and build new landscapes that ingeniously capture notions of place.

Marx evocatively wrote "all that is solid melts into air", regarding the creative and destructive power of capitalism. The phrase rings true today for completely different reasons, in particular how all our accepted orthodoxies have been thrown out the window. How our once solid and understood environment has fundamentally melted, both environmentally and economically.

As landscape architects we are caught in a bind, do we hold dearly to our past convictions whilst the ground is shifting rapidly beneath us, or can we instead reverse 30 years of complacency to a more activist role?

If we speculate for a moment and describe who might be an ideal candidate to deal with our constructed environments, in these challenging times, it might read something like this: Someone who can think outside the box, able to understand the world at the macro scale, a person that can facilitate and choreograph multi-faceted disciplines, and offer creative and sustainable solutions that bridge between science, environment, ecology, culture and art. I might be a bit biased, but that candidate sounds to me like a landscape architect. To put it bluntly, as landscape architects, we can either respond meekly to the circumstances of which we are confronted, become bystanders to the main event, or instead utilize our distinct skills to be part of the solution.

In Australia, the need for some design bravado, to contribute meaningfully to our unique environment, is more relevant than ever. It is a continent where landscape architects can creatively engage in an amazing diversity of scales, environments and extremes.

Yet our dissolving environmental conditions, diminishing water, parched summers and increasingly dry winters, contrasted with a suburban mentality of sprawl demands designers that can create alternate solutions within a new paradigm. This does not necessarily require a new manifesto but instead a careful repositioning to ensure what we do best offers alternate and viable solutions for our constructed environment. Thankfully this is already happening, albeit I suspect, somewhat too quietly.

In our Australian context successful projects of this new paradigm challenge the orthodox brief and advocate an alternate position, demonstrating how one might offer truly sustainable outcomes, yet imbue the solution with unexpected, joyous and rewarding qualities.

Successful projects are borne out of an enlightened policy framework that encompasses an all of government approach to a sustainable future. Successful projects no longer try to import distant ideas but instead demonstrate an ability to conceive of and build new landscapes that are derived of their place, and ingeniously capture the specificity of its culture and environment and all their nuances. Successful projects no longer cloak our built environments with superficial greenery but instead demonstrate our skills in design leadership, choreographing diverse disciplines to create truly integrated outcomes encompassing, architecture, ecology, infrastructure and artistic practice. Successful projects instill a sense of the avant-garde, demonstrating possible alternate futures and communicating boldly their role in the world.

Most importantly we must find a means to permeate our world with a sense of hope, and humanity. Creating spaces and environments that are sustainably driven is admirable. Capturing the hearts and emotions of those that inhabit them will be the more tangible measure of success.

皇家墨尔本理工大学建筑与设计学院副教授
B.L.Arch MD（城市设计）澳大利亚景观设计师协会准会员
Taylor Cullity Lethlean 总负责人（澳大利亚墨尔本，阿德莱德）
www.tcl.net.au

出色的设计彰显出超凡的想像力，并会因地制宜地营造新的景观。

"一切坚固的东西最终归于烟消云散"，马克思曾用这句话来形象地描述资本主义的创造力和破坏力。这句话用在当代很多地方也是非常合适的，最明显的例子就是那些曾坚信不疑的陈规最终都不复存在了。我们熟悉的软环境和硬环境曾经是多么的坚不可摧，如今也在慢慢地改变着。

景观设计师陷入了困境，环境迅速变化时我们是否执着于陈规？我们能否丢掉 30 年来的成就去转换角色？

如果我们静静地思考一会儿，就会发现谁能在这个充满挑战的时代里担起环境改造的重担，能够跳出思维定式，从宏观角度来解读世界，并能够制定多方位的规划宗旨，提出具有创意且可持续发展的解决方案，将科学、环境、生态、文化与艺术完美地结合在一起。在我看来这样的人选只有景观设计师，也许这样说会比较片面，但景观设计师可以从一个旁观者变成主事者，改变所处的环境，也可以各显其能为设计方案贡献一份力量。

在澳大利亚，依据独特的环境做出突破性的设计显得尤为重要。景观设计师可以在这片土地上将自己的创意应用于各种不同的项目中。

然而，面对日益恶化的环境、水资源逐渐枯竭、夏天越来越炎热、冬天越来越干燥，而郊区发展模式为景观设计师提供了一种新的解决方案。这并不需要发表新宣言，但需认真地重新定位，以设计出可行方案来改造建筑环境。庆幸的是，这样的改变已经悄然开始。

在澳大利亚，运用新模式的成功项目颠覆了固有的思维，挑战着传统的思想，向人们证明在确保环境可持续发展的同时也能够设计出高水准的方案。

一个成功的项目离不开政府的支持；一个成功的项目不再是移植外来的创意，而是要彰显自身的创造力并打造源于本土的景观设计，巧妙地捕捉到独特的本地文化、环境以及其微妙的关系；一个成功的项目不再依靠表层绿化装扮环境，而是运用设计技巧和各种方法创造一个集建筑、生态、基础设施、艺术于一体的综合方案。一个成功的项目蕴含着先锋思想，改变着未来，向世界展示着其重要地位。

创造可持续发展的空间与环境值得提倡，而捕捉其内在的精髓才是最大的成功。最重要的是，我们必须要通过设计将希望与人性播撒到世界各地。

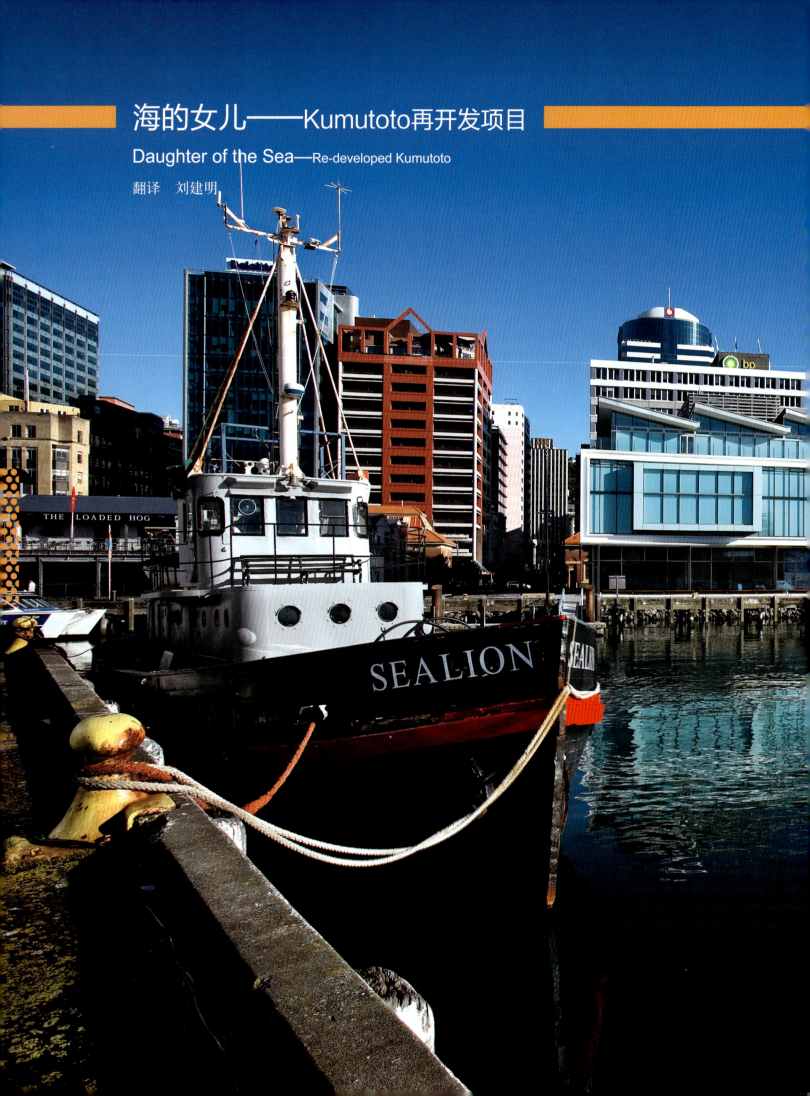

海的女儿——Kumutoto再开发项目

Daughter of the Sea——Re-developed Kumutoto

翻译　刘建明

依偎在惠灵顿港的天然怀抱中的新西兰首都惠灵顿是一个多次从地震灾难中复原重振的城市，其横跨活跃的地震断裂带，常年接受海风的洗礼。而 Kumutoto 则是惠灵顿滨水地区最近重新被开发的一块区域，项目竣工后，惠灵顿将再次向大海敞开怀抱。

惠灵顿滨水有限公司首先与 Studio Pacific Architecture 商讨了项目的总体规划，尔后又与 Isthmus Group/Studio Pacific Architecture 共同筹划创建新的城市空间。Kumutoto 长久以来一直被用做停车场，但经常会有多余的空闲车位。Kumutoto 是以历史上的一个村庄 "pa" 的名字来命名的（"pa" 是土著毛利人的村庄或定居点的名称），在改造后的地下管路中依然流淌着一条很早以前就存在的小溪。Kumutoto 的改造方案每一处都淋漓尽致地体现了美学理念，并且与城市的悠久历史、气候、景致和背景非常契合。设计团队极力突出本案的几大关键特色：码头 "步行道"、Kumutoto 小溪、城市网格和港口。

通过新建的固定式码头和浮桥，扩展了码头广场的外延。照明设备、高低错落的坐椅和桃金娘科常绿树种在城市和港口之间交替排开，为规划布局和动态景观增添了多样化色彩。旧码头的入口现在已对外开放，增强了通达性和游客的流动性。

在 Kumutoto 广场，重新引入的溪水令小溪再度绽放活力，停车场被改造成河道，并且重新开设了一个入海口。沿着溪岸改造出一系列不同的地形，人们不仅可以在此避风，还可以更亲密地与溪水互动。在新建的具有零售和商业用途的建筑物后面，隐藏着一处包罗万象的城市空间，这里有巨大的漂浮的原木、混凝土坐椅以及年久失修的港务局遗迹。从广场上可以观赏到落日的余晖，建筑物巧妙的布局设计使得咖啡馆和饭店呈点状散布，使整个空间充满活力。

户外设施的设计更突显了开放公共空间的这一特色。整套户外设施既彰显 Kumutoto 的独特风格，又通过选材、颜色和尺寸等方面来保持与其他滨水区域的紧密联系。桥梁、灯柱、木排坐椅和溪流入海口等各色设计元素，构建成一处彼此牵连、浓情四溢的滨海景观。

由于滨海地区缺少植被，因此植物配置的多样性受到了很大的局限，这就使得单一物种大量繁殖，将地平面割裂

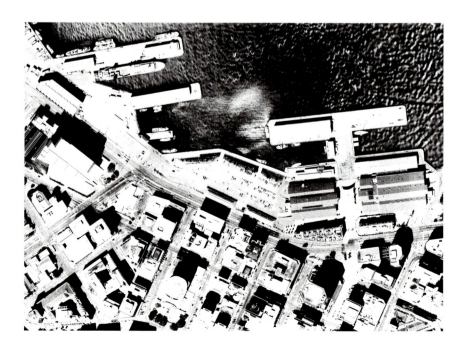

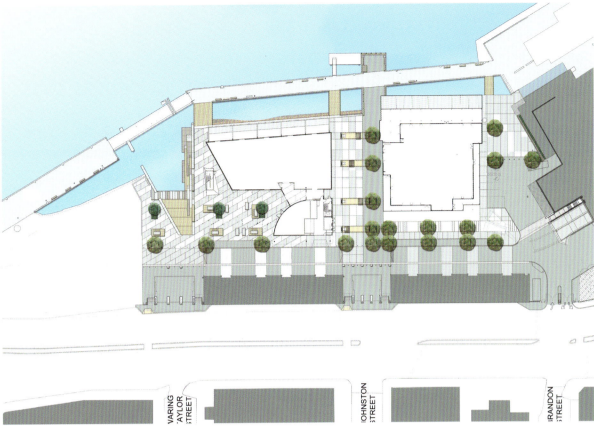

成单一物种独占的多个区块。历史悠久的防波堤附近是金边
剑麻的专属领地，让人们不禁忆起这里曾经是盛极一时的亚
麻交易场地。曾经生长在 Kumutoto 广场的 Apodasmia similes
在溪流沿岸的潮湿沙丘中繁衍，并沿着惠灵顿海岸大片生长。
此外，还有土生土长的棕竹和悬铃木为盛夏的广场提供了更
多的阴凉区域。

　　地面铺设材料以纹理和坚实程度作为选取标准，使用
聚合混凝土和打磨的混凝土面板来打造一种类似织锦的效
果，鳞状薄片和码头原木的使用则可以保留该地区原本坎坷
不平的风貌。在新修通道的沿线旁边，巨大的镶边石很容易
让人联想起历史悠久的 11 号、13 号屋棚附近的古老的防波
堤，而最初用来搭筑屋棚的那些历史久远的原木被巧妙地循

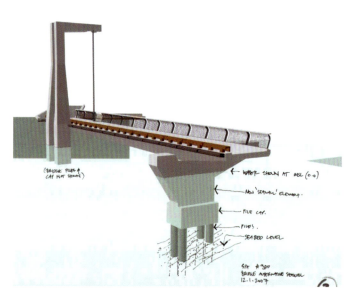

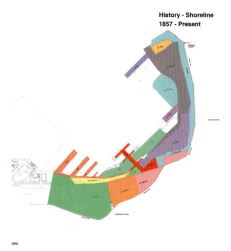

环再利用，重新铺筑在码头广场上，给广场的地面平添了一份暖意。

历史悠久的拖船码头被精心修缮，以更好地突出厚重的历史传承。在靠近Kumutoto出海口有一座新建的雕刻步行桥，仿佛是漫长、延伸的步行道的一个"定锚点"。步行桥增强了出海口和港口之间的视觉联系，桥的一端依托一个"摇篮"状结构来支撑，另一端却是一个"塔式起重机"，将桥上的人流和其他各种信号完全展现出来，吸引人们前往滨水地区游览。

从开放的那一天起，惠灵顿市民就将Kumutoto作为他们生活中的一部分，他们不惜主动改变早已习惯的行走路线，而将阴凉舒适的广场作为暂别繁华都市的栖息地，偷得一刻闲暇与港口倾心互动。

Straddling active earthquake fault lines and blasted by ocean winds New Zealand's capital is a resilient city nestled in the natural amphitheatre of Wellington Harbour. Kumutoto is the latest piece of the Wellington's waterfront to be re-developed, reorientating the city back towards the sea.

Named after a former pa (a "pa" is a village or settlement of the indigenous Maori people) and an ancient stream running in a culvert under the site's reclaimed land, Kumutoto had become a redundant carpark until Wellington Waterfront Ltd. firstly engaged Studio Pacific Architecture to masterplan the site and then later engaged Studio Pacific Architecture with Isthmus Group to create new urban spaces. The result is characterised by a robust aesthetic echoing the site's rich history, climate, views and connections. The design drew on the site's key characteristics: the wharf "promenade", the Kumutoto stream, the city grid, and the harbour.

Wharf Plaza seeks to reinforce the city grid, extending the reach of the city with a new fixed wharf and a floating pontoon. A sequence of lighting structures, rugged seats and Pohutukawa (Metrosideros excelsa) create a procession between the city and the harbour, creating a variety of opportunities for occupation and movement. Heritage wharf gates are now fixed open, enhancing movement and access.

At Kumutoto Plaza itself, the stream is revealed and celebrated by pulling the edge of the water back, declaiming the former carpark and creating a stream mouth. A series of terraces spill down to the water's edge, allowing people to engage more closely with the water and sheltering from wind. Tucked behind a new retail and commercial building, a more contained urban

space features a series of large, floating timber and concrete seats reminiscent of Harbour Board skids. This plaza captures the afternoon sun and through careful building positioning allows the surrounding cafés and restaurants to spill out and activate the space.

Outdoors furniture further defines open spaces. A set of furniture was developed that was at once specific to Kumutoto

while still maintaining links to other waterfront areas through material, colour and scale. Formal relationships create a family link between distinct elements such as the bridge, the light poles, the raft seats and the stream mouth.

With no existing vegetation on the site, the plant palette is deliberately minimal, allowing single species to colonize zones cut from the ground plane. Adjacent to the historic seawall Phormium tenax is introduced in restricted blocks, bringing the flax back to an area where its trade was so influential. Apodasmia similes, used in the Kumutoto plaza, is naturally found in the damp sand dunes surrounding streams as they emerge along the Wellington coast. Further structure is given to the area by native Pohutukawas (Metrosideros excelsa) while Oriental Planes (Platanus orientalis) add summer shade to the more sheltered plaza.

The surface materials were selected for their texture and robustness, with exposed aggregate concrete and honed concrete panels used to create an evocative tapestry, while the scale and use of traditional wharf timbers ensures that the area still keeps its rugged grain. Large kerb-stones along the new

laneway make reference to the old seawall adjacent the historic waterfront Sheds 11 and 13, while the historic timber that once was set surrounded the sheds were carefully recycled and re-laid in Wharf Plaza, adding texture and warmth to the ground surface.

The historic tug-boat wharf was renovated to express its rich heritage character, and adjacent to the Kumutoto stream mouth a new sculptural pedestrian bridge was created acting as an anchor point along the extensive promenade. The span of the bridge deck span acts to strengthen the visual connection between stream mouth and harbour and is supported at one end by the "cradle", and at the other the "tower crane" which frames movement and beckons to those on Wellington's main street, drawing them to the waterfront.

From the day it opened Wellingtonians have embraced Kumutoto, as part of their lives by actively changing their commuting routes, using the sheltered terraces as a respite from the lively city streets and to simply pause and engage with the harbour.

项目位置：新西兰惠灵顿 Kumutoto
客　　户：惠灵顿滨水有限公司
预　　算：约 650 万美圆
占地面积：12 000 m²
项目时间：2002 年~ 2008 年
建筑设计：Studio Pacific Architecture
景观设计：Isthmus Group
所获奖项：2008 年新西兰景观设计师协会（NZILA）
　　　　　最高成就奖
　　　　　2008 年新西兰景观设计师协会（NZILA）
　　　　　城市设计金奖
　　　　　2008 年新西兰建筑师协会（NZIA）本地奖
　　　　　2008 年新西兰承包商联盟建筑奖
　　　　　2008 年惠灵顿／怀拉拉帕承包商联盟奖
　　　　　2008 年 IES 灯光设计优胜奖

Location: Kumutoto, Wellington, New Zealand.

Client: Wellington Waterfront Ltd.

Budget: approx. $6.5 million

Site Size: 12,000 sqm

Project Dates: 2002~2008

Architect: Studio Pacific Architecture

Landscape Architect: Isthmus Group

Awards: Winner NZILA (New Zealand Institute of Landscape Architects) Supreme
Award, 2008

Winner NZILA (New Zealand Institute of Landscape Architects) Urban
Design Gold Award, 2008

Winner NZIA (New Zealand Institute of Architects) Local Award, 2008

Winner New Zealand Contractors Federation Construction Award, 2008

Winner Wellington/Wairarapa Contractors Federation Award, 2008

IES Lighting Awards - Award of Excellence, 2008

悉尼大学达灵顿校区主路景观设计

Landscape Design for Darlington Public Domain, University of Sydney

翻译　王玲

　　该项目主要的设计目标是可以使所有行人都拥有同等的道路权，其中包括从牧羊人大街入口前往新主楼的。此外，景观规划还着眼于在环境和社会两个方面来提升达灵顿校区的环境品质。雨水收集有助于湿地的形成，湿地收集储存的雨水又可以重新用于灌溉花园和草坪。设计师在校园里创建了许多新的公共空间，并设计一个新的分类花园来扩充原来的花园。

　　该设计使校园的主入口焕然一新。校园入口的雨水花园与一个将主步道引入校园的结构相连接，这种设计有效地解决了现有道路通达性问题和公共安全的问题。设计师还通过横贯校园的木质人行桥解决校内道路导示问题，木质人行桥是一个清晰定义出入口的大胆的城市景观尝试。设计师与保罗·卡特通力合作，共同确立以场地上的欧洲和土著文化历史为基础的艺术策略和设计理念；这样也为有效融入设计结构中的艺术作品确立出了框架。"金色树林"策略成功地将各种设计元素融入到一个没有任何艺术预算的项目中，艺术作品包括印在人行桥金属外立面上的诗歌。木质人行桥的地面上镶嵌了 3000 支采用低耗能荧光管的亚克力灯，灯光由下向上漫射出来。"金色树林"策略体现出沿主路的灯光图案和种植区的形态。客户对牧羊人大街入口设计的最初设想是为每天经由此处进入校园的 20 000 名学生提供一个公平安全的接入点。设计师在满足客户要求的同时，打造出一个令人难忘的校园入口，不仅转变了该地区的工业形象，更体现出人们希望改善学校正门前往 Redfern 区的环境愿望。

　　设计师利用清晰明了、照明良好的校园交通动线来满足和增强校园安全感的社会要求。设计通过"金色树林"的策略来诠释场地的文化和历史特色。在道路设计上顺应场地的地形变化，并恢复校内 Blackwattle 河支流的生态系统，使之

具有较高的品质。人行路桥采用镀锌钢框架结构，框架嵌在原有的混凝土墙体之上。桥体一侧的一排钢片不仅吸引了人们的注意力，而且还在适应环境的情况下巧妙地将周围的工业建筑掩映起来。扶手和栏杆都采用木材和不锈钢进行细部处理，一条直径2.6cm的工业用不锈钢缆索栏杆将桥上的钢片串联起来。

　　该项目的环境工程主要体现在处理达灵顿校园内雨水的集水战略上。雨水流经生物滞留系统，然后储存在一个10万升的地下蓄水池中。蓄水池中的水可以满足校园内所有植被的灌溉需求，多余的雨水则继续流入Blackwattle河，水经过生物处理后对下游流域具有积极影响。校园门口的生物滞留雨水花园是场地设计的一大亮点，也是学校鲜活的教育和宣传范本。花园利用悉尼当地植被有效地促进了植物分类学的教育，这不仅支持了大学的教学规划，同时也推动了该地区城市生态学的发展。

　　该项目在现代独特的设计方案中不仅实现了它的环境目标，同时也呼应了场地的自然历史和社会文化背景。它是一个由景观设计师带领完成的成功典范，一个解决了多项工程问题的、大胆的城市设计解决方案。此外，该项目并没有因为过度工程化而削弱其影响。其成功地将竞标阶段的理论观点付诸实施，并与学校各部门进行长期有效的沟通，最终实现了设计所要承载的广泛物质需求、多种价值工程和成本的削减，是一个在竞标意向中就直接体现出其设计成果的杰出代表。

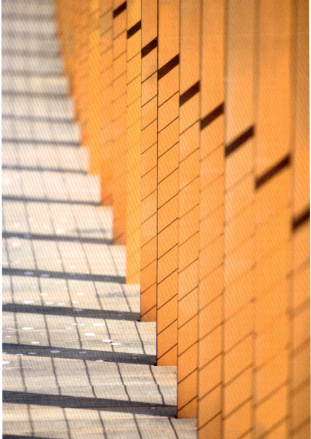

The primary aim of the Darlington Public Domain project is to provide equal and equitable access for all pedestrians to travel to and from the Shepherd Street entrance to the new USYD Central Building. In addition to this, the landscape scheme looks to improve the environment of the Darlington campus both environmentally and socially. Water harvesting of the local stormwater catchments aims to create wetlands that treat and store water for reuse such as irrigation supply to gardens and lawn. Creating new public spaces within the campus and the development of a new taxonomy garden is regarded as a replacement and expansion of the existing garden.

The design transforms the image and identity at a key entry to the University. It integrates a bio-retention water garden at the entry of the University bridged by a structure that carries the main pedestrian access into the campus. Whilst resolving existing equal access and public safety issues. The design resolves way-finding issues within the Campus by extending the timber decking of the bridge as a pedestrian path extends through the Campus, as a bold urban element that clearly defines entry and departure. The designers collaborated with Paul Carter to establish an art strategy and design philosophy that is based on both the European and Indigenous cultural histories of the site. This set up a framework for artwork to be meaningfully integrated into the fabric of the design. The Golden Grove strategy has been successful in the infusion of varies elements into a project where no art budget was allocated. The artworks include poetic texts stenciled into the steel façade of the bridge. There is an array of 3000 lights in the timber deck, created by drilling holes and backfilling with

acrylic, in addition to uplighting from beneath with simple low energy fluorescent fittings. The Golden Grove strategy informs the lighting pattern and the shape of the planting areas along the main path. The primary aspect of the client's brief for the Shepherd Street entrance was to create a safe and equitable access point to the university for the 20,000 students that use this each day. Our response has achieved the clients brief and also created an entrance to the University that is memorable, transforming the industrial character of the area and expresses the environmental aspirations of the University at its front door to Redfern.

The design responds to the social requirements of improved safety on the Campus by creating a clear, defined and well-lit line of travel in and out of the campus. It responds to the cultural histories of the site through the Golden Grove Strategy. And it engages with the topography of the site in resolving access issues whilst reactivating the ecological systems of the Blackwattle creek tributaries within the campus. The resulting quality and finish of the build project has been of a high quality. The bridge is constructed of a galvanized structural steel frame sitting on an insitu off form concrete wall with a façade of steel blades that lead you into the campus and screen views of the adjacent industrial builds yet respond to their context. The handrail and balustrade details have been carefully detailed in timber and stainless steel with the unique feature of a top rail running through the steel blades being made from an industrial 26mm diameter SS cable.

The project's environmental engineering was conceived to develop a water harvesting strategy that treats stormwater run

off from Darlington. The water passes through the bio-retention system and is then stored in a 100,000liter underground tank. This is designed to supply all the Darlington campus irrigation requirements. Excess water from the treatment system continues on to Blackwattle Bay. The water polishing actions on campus will have a positive impact down stream from the campus. The bio-retention garden at the entrance of the University became a feature of the site, and an educational and promotional asset for the institution.

The garden areas were designed to assist in the education of Taxonomy, using local Sydney flora. This is a design gesture that supports the University's education program and is supportive of the urban ecology of the local region.

The project achieves environmental goals in a contemporary and unique design solution, while also being responsive to the site physical, historical, cultural and social context. It is a good example of landscape architects taking the lead of a team of engineers to deliver a bold urban design solution consisting of many engineering issues without being diluted by potentially over-engineered outcomes. This project has been successful in carrying the theoretical ideas of the competition stage through; long consultation process wit the campus facilities, the vast physical requirements that the design had to accommodate, the multiple value engineering and cost cutting reviews, to deliver an strong design outcome that is still directly informed by the competition intent.

项目位置：澳大利亚悉尼市
客　　户：悉尼大学
预　　算：200 万美圆（280 万澳圆）
建成时间：2007 年（一期）；
　　　　　2009 年（二期）
景观设计：Taylor Cullity Lethlean 与
　　　　　Paul Carter
艺术顾问：Paul Carter

Location: Sydney, Australia
Client: University of Sydney
Budget: US$2 million (AU$2.8 million)
Completed Time: 2007 (Stage 1);
　　　　　　　　　2009 (Stage 2)
Landscape Architects: Taylor Cullity Lethlean
　　　　　　　　　　　　with Paul Carter
Artist: Paul Carter

奥克兰布朗斯湾改造
Regeneration of Browns Bay, Auckland

翻译　李沐菲

布朗斯湾位于奥克兰北岸的东海岸，与许多其他海岸城市一样，布朗斯湾因不断改造而焕然一新。由北岸市议会提议，设计师旨在通过设计一种能使市中心和海滩都临街的巷道来恢复城市街区的作用。这些巷道的建立可以加强城镇和海滨区的联系，而这些跨海滨巷道空间开放后可以创造新的步行道和广场。

新的巷道被设想为一个公共的空间，一个"允许汽车通行的步道"，因为在这个空间里行人具有优先权——在联系了城镇与海滨资源的同时，还保存了双向汽车通道和停车场。这个设计目标是通过使用一个狭窄的通道来实现的，而这个狭窄的通道通过提升行人广场的高程来创造一个低速(30kph)的环境。

该项目是由强有力的硬质木板、基座和混凝土巷道组成，微妙的组合与海滨环境遥相呼应，更加突显出海滩前方巷道的杰出设计。所有的设计元素均反映了滨海地区的自然景观，采用一系列新开发的基座，所有的材料都是精挑细选的，而且保证在海洋气候下不会变质，例如硬质木材、不锈钢。应未来节日活动的需要，相应的法规也会出台，如引进先进的技术和建立更多的服务站点。

通过安装路灯来保证可持续照明，这些路灯排放的二氧化碳较少，减少了能源消耗，具有耐用的特点。同时，长廊采用透水砖并配有排水过滤系统，以消除来自城市的垃圾和其他污染物，这一点已成为巷道设计的一大特点，既提高了水的质量，又确保了不让污染物流入大海。

该项目保留了非自然形式的现存沙滩保护区，以及木质巷道两侧现存的草坪和高大树木，这样既能保留最初的设计目标，也能表现出整个设计理念的推进过程，以此来弥补现场的局限。木板保护着滴灌管道，而管道包围着树木，以确保树根吸收充足的水分。

从初步设计到最后与政府签订正式合作协议，标志着该项目的成功开展。海滩前方巷道的竣工标志着后街被改造成一个具有吸引力的、受行人喜欢的长廊。这里不仅是当地人的最爱，也将成为远道而来的游客们所喜爱的旅游地。

Located on the east coast bays on Auckland's north shore, Browns Bay had like many other coastal towns turned its back on its beachfront as it developed. Engaged by North Shore City Council, Isthmus aimed to reactivate the town's urban blocks by means of a new laneway giving both a frontage to the town centre and foreshore. New connections through the blocks proposed to improve connectivity between the town and the beach reserve. Where these cross the Beachfront Lane spaces open out to create new pedestrian zones and plazas.

The new laneway, conceived as a shared space in which pedestrians are to have priority over vehicles is designed as a "promenade with cars", linking the town with the beach reserve whilst maintaining two-way vehicle traffic and parking. This has been achieved by utilizing a narrow carriageway that rises over pedestrian plazas to achieve a low speed (30kph) environment.

Consisting of a robust hardwood timber boardwalk and seating and a concrete laneway, the subtle palette of materials was adapted to fit in with and enhance Beach Front Lane's outstanding coastal setting. The furniture elements were designed to reflect the nature of the site's beachfront with a range of seating options developed and detailed in materials typically found in a marine environment such as hardwood

timber and stainless steel. Provision for future festival events was accommodated through the introduction of power and service outlets.

Environmentally sustainable lighting solutions were achieved through the installation of lamps that emit less CO_2, use less energy and have a longer life. The use of permeable paving along the promenade together with a drainage filtration system to remove litter and other pollutants from urban runoff, has also become a feature of the lane that improves water quality and ensures that polluted runoff does not flow into the sea.

The informal nature of the existing beach reserve with lawn and mature trees is maintained with the timber boardwalk providing an example of how the concept design evolved to meet the constraints of the site whilst retaining the initial design objectives. Trees were protected by raising the boardwalk onto piles to clear the roots within the dripline.

From the early stages of the concept design through to the formal opening ceremony collaboration with the community was a key factor in the project's success. With the completion of Beach Front Lane, a back-street has been transformed into an attractive pedestrian-friendly promenade that has become a very popular destination for locals and people from further afield.

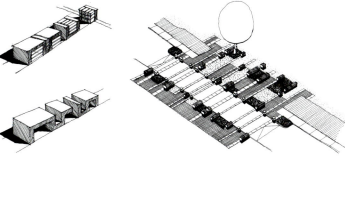
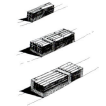

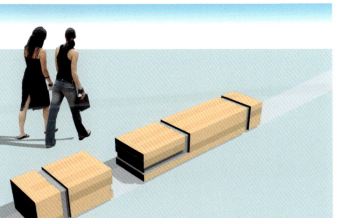
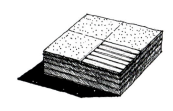

项目位置：新西兰奥克兰市布朗斯湾

景观设计：城市地峡设计

客　　户：北岸市议会

占地面积：12 500m²

预　　算：180 万美圆

项目时间：2005 年 ~2008 年

荣获奖项：2007 年 ~2008 年新西兰公路建设最高荣誉奖

2007 年 ~2008 IESANZ 照明设计杰出奖

Location: Browns Bay, Auckland, New Zealand

Landscape Architecture: Isthmus Urban Design Landscape Architecture

Client: North Shore City Council

Surface Area: 12,500 sqm

Budget: US$1.8 million

Project Dates: 2005~2008

Awards: Highly Commended Roading NZ 2007/2008

IESANZ Lighting Design Award of Excellence 2007/2008

乡土景观——Eynesbury城市预留地

Rural Landscape—Eynesbury Township

翻译 王玲

Eynesbury 是一处距墨尔本以西 40km 的城市预留地，位于快速发展的华勒比和梅尔顿之间。它也是该地区最大的一块富有历史意义的独立产权地块。

该地区拥有 290 000 户居民。如今，正逐步发展成一处集环境保护与娱乐休闲于一体的"传统邻里型"的综合住宅区。凭借悠久的历史沉淀，该地区最终将吸引超过 10 000 户居民来此定居，而预留地的中心也将满足各种商业零售、教育以及文娱设施的功能需求。最引人注目的是 72 850m² 的原始地块中仅有 16% 被重新划分后用于开发建设，其余地块则是在预留地周围的农业耕地、天然草场、Grey Box 森林以及河流廊道。

该项目致力于营造强烈的社区认同感和归属感，尊重各种不同"澳式"乡土生活的价值观。项目早期开发主要集中在高尔夫锦标赛的球场建设和恢复昔日 Eynesbury 中心的一座庄园。林草覆盖的高尔夫球场与一行行干草交相辉映，蜿蜒逶迤的公共道路和开阔的空间系统点缀其间，巧妙地将未来住宅与原来的庄园连接在一起。

修复后的庄园如今已改建为会所，内设有各类公共设施，而且它将与社区内建筑及其相关的娱乐健身设施一样可以得到不断的改善。公共道路系统将未来所有的住宅区与公共设

施和镇中心连接在一起，并鼓励人们将步行和骑自行车作为日常生活方式的一部分。

该地区的规划是基于一种正交网格模式，围绕一条主要街道展现"村镇"的自然风貌。为了扩建和活跃这条主要街道，学校等大型设施建在半径为 400m 的范围处。这样便可以在不增加密度和不破坏预留地的情况下，使人们步行便可到达学校。预留地的规划也体现出诸多可持续性特点，这也正是该地区在未来需要关注的方面。

该地区规划的重点是在已有丰富的自然环境中，展现其环保特性。许多区域重新种植了本土植被和树木，以此来拓展该地区的生境长廊。本地的自然和再生材料，如石材、木材和铁被广泛地运用在项目的硬质景观设计中，展现出该地区所特有的乡土气息。

2007 年预留地一期开始建设，同年高尔夫球场和会所投入使用；早期的住宅规划也于 2008 年 9 月竣工。自然环境和文化景观的保护与提升是该地区总体规划的灵魂所在，无论是游客还是未来的居民都会对这里的景观深怀感激，因为它不仅独具匠心，同时还营造出独特的生活环境。

施工范围

紫锥菊　杜松
雪美人鱼
阔叶麦冬与白百合
阔叶麦冬与白百合
屈曲花与白烛葵
杜松
雪美人鱼
区域B
边角植物
雪美人鱼
阔叶麦冬与白百合
A区的中国木槿
红缬草
雪美人鱼
新西兰麻
墨西哥白鼠尾草
英国薰衣草　区域A
天堂鸟
阔叶麦冬和白百合
天堂鸟
海边雏菊
屈曲花与白烛菊
区域C
印度山楂
香雪兰
蓝蓟属
英国薰衣草
杜松
印度山楂
区域D
宝血草
非洲菊
紫锥菊
区域E
象耳——大叶植物
新西兰麻
婆婆纳
窄叶朱蕉
金银花　区域P
水芋
蝴蝶灌木
巨朱蕉
区域S
五叶地锦
月桂树
紫藤
区域Q
蔬菜园
黄杨
君子兰
芦荟藜
百合花
区域F
区域M
英国薰衣草
伏牛花
区域N
玫瑰公园
巨朱蕉
区域O
芦荟
区域H
冰叶日中花
无刺丝兰
红川莲
英国薰衣草
金缕梅属
百合花
石莲花
火炬花
尾龙舌兰
紫万年青
冬日玫瑰

SCALE 1:200@A0

英国薰衣草
天堂鸟
杜松
君子兰
火炬花
紫薇科树
白菊
五叶地锦
Canna 百合花
红缬草
巨朱蕉
尾龙舌兰
蓝蓟属

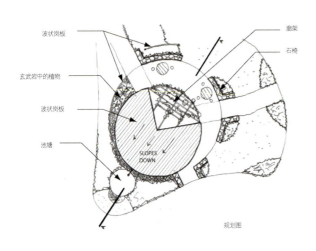

規劃圖

波状岗板
玄武岩中的植物
波状岗板
池塘

廊架
石椅

SLOPES DOWN

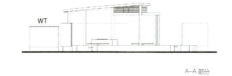

WT

A—A 部分

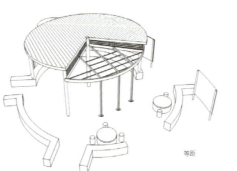

等距

攀缘紫藤

Eynesbury is a new township located 40 kms west of Melbourne, between the growth areas of Werribee and Melton. The historic station property is the largest parcel of land in the one ownership, in the region.

The unique town of 2900 hundred homes is being developed as a integrated residential community in a "traditional neighbourhood style", within an environmental and recreational setting. Building upon the former station property's historic past, the township, will eventually be home to over 10,000 residents, who will benefit from a town centre that will support civic, retail, commercial, education and recreation opportunities. Of note, is that only 16% of the 7,285ha of original property has been rezoned for development. The remainder will continue as farming operations and protected native grasslands, Grey Box forest and river corridor, surrounding the new township.

The Eynesbury Township seeks to provide a strong sense of community, ownership and belonging and is based on many of the values of "Australian" country living. Early phases of the township are centred on a Championship Golf Course and the restored station Homestead; the original heart of Eynesbury. The grassy woodland-style golf course is set amongst existing windrows and features a generous shared path and open space system that services future residential stages with linkages back to the homestead facilities.

The restored Homestead is now home to the Club house and amenities and will be enhanced by a purpose-built community building and associated recreation and sporting facilities. A shared path system will connect all future neighbourhoods with these facilities, and the town centre, to encourage walking and cycling as a daily part of life.

The town itself is based on an orthogonal grid pattern to reflect the "country town" nature of the setting, around a main street principle. To engage and enliven the main street, the larger uses like the school were placed at the periphery of the 400m radius so that it was in walking distance, but did not compromise the density and use of the township itself. The township plan addresses many of the sustainability issues that are of concern within the community in the new millennium.

Emphasis has been placed upon creating a strong environmental focus within the already rich setting. Many areas have been revegetated using indigenous plants and trees, to extend habitat corridors across the property. The use of site sourced, natural and recycled materials such as stone, timber and iron found have been used extensively in the hard landscape, to reflect rural sense of place at Eynesbury.

2007 saw the first stages of the new town commence construction, with the opening of the Golf Course and Clubhouse facilities and early residential stages were completed in 2008/2009. Tract continues to provide planning and landscape architectural services for this innovative township, translating the overall masterplan into a successful on-ground development. The preservation and enhancement of the natural environment and cultural landscape is inherent in the masterplan for the township. Visitors to the site and future residents will gain a tangible appreciation of the role that the landscape plays in providing a unique living environment, hat sets it apart from other developments in the region.

项目位置：澳大利亚维多利亚州墨尔本市西南 40km 处

占地面积：7285 万平方米

客　　户：Eynesbury 合资发展集团（Geo 地产集团和 Woodhouse Pastoral 公司）

预　　算：约 2000 万澳圆

建设时间：2007 年～2009 年

景观设计和城市规划：Tract 咨询公司

城市设计：BDA 建筑设计事务所

高尔夫球场设计：Graham Marsh 高尔夫球场设计

所获奖项：澳大利亚景观设计师协会（AILA）维多利亚州优秀奖

Location: 40km south west of Melbourne Victoria Australia

Site Size: 7,285 ha.

Client: Eynesbury Joint Venture Development (Geo Property Group and Woodhouse Pastoral Co.)

Budget: opprox.AU$20 million

Project Dates: 2007 ~ 2009

Landscape Design and Town Planning: Tract Consultants

Urban Design: BDA Architecture

Golf Course Design: Graham Marsh Golf Design

Awards: Australian Institute of Landscape Architects (AILA) (Merit Award) VIC

炫丽的Gogo之舞——城市花园景观

Toorak a Gogo—City Gardenscape

翻译　刘建明

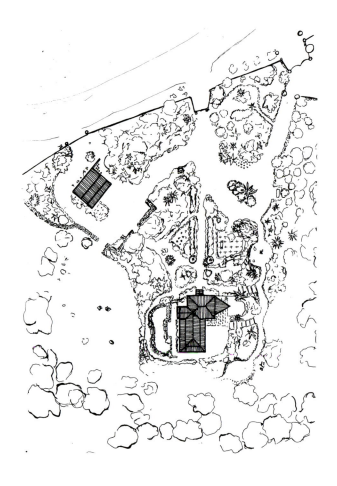

焦黄的草丛掩盖着锥形的火山灰，青石黑土上星星点点的矗立着几棵灰色的木麻黄树，满是黄褐色的野草和绵长的山楂树篱，还有品尽人间沧桑的榆树林阴道——这就是澳大利亚维多利亚州西区诺亚特山区（Noorat）的代表性景观，很显然这幅画面与其极力想要再现的英国风情相去甚远，毕竟这儿并不是英格兰。

沿着斜坡踏着火山灰爬至半山腰，人们会看到丛生的树林——望不到边的橡树和澳洲木麻黄，威廉·马丁放弃了最能表现英伦风情的 Rosemary Verey、Gertrude Jekyll 以及晚期得到 Gertrude Jekyll 真传的澳洲后辈 Edna Walling 的风格流派，决然以本土的橡树和木麻黄来表现故乡景观的精髓。设计师并不担心该项目最终会演变成巴厘岛、日式花园或文艺复兴时期花园的景观风格，因为这里没有浪漫，没有绿色，也没有郁郁葱葱的夏日田园风情。设计师以其对这块土地的了解和体验作为创意的起点：夏日的干燥、冬日的潮湿、焦黄的枯草、嫩绿的草场和亘古不变的青石黑土；在无垠的蓝天和多变的云朵下面，狂野的海风终日在这块土地上肆虐；山坡上的自然风情，还有那远处原汁原味的远古景观。基于这样的理解，设计师运用个性化的手法来打造一个本土花园，并倾尽全力设计一个打破"丛林中的简单生活"的陈腐理念的花园。维格迪亚是一个花园，它随着季节的变化而呈现出不同的景观。设计师将花园的韵律谱入一段音乐剧中，那是一曲滑稽歌剧——婉转轻柔、机智诙谐、暗藏险峻、略显张扬。急板——短暂的冬日与春天的鲜花；强音——高耸的 furcreas 和 dasylirions（稠丝兰属植物）；钢琴——羽毛状的青草；慢板——浅浅的橘色、红色、棕色、灰色植物形成的波浪；拨奏——长而尖的芦荟和龙舌兰；二二拍——具有短暂、急促表现效果的十字木和纳塔尔梅花；渐速音——高挑、泛红的醉鱼草，大片的芦苇和 melianthus；渐强乐段（高潮）——压倒性的景观；广板——夕阳渐渐没入西山，暮色笼罩着花园。乐章和整个歌剧的创作都是根据设计师对每一种植物和气候"乐器"精妙的观察，用娴熟独到的演奏技巧一气呵成。每一个音符都是如此的和谐完美，足以激发新奇的理念和美妙的联想；雕塑和固定设施给花园增添了活力和锐气，整个设计在创作过程中灵感频现。每一个音节和旋律配合都能够带来现实的震撼效果，让人们深切地感受到此处的惟一性——只此一地，别无他处。正如 Philip Glass 和 Ian Hamilton-Finlay 的传世精品一样，均是旷世之作。

A volcanic cinder cone covered in seer yellow grass and a few dusty Casuarina trees rises above a sparse landscape of black soil, black rocks, more yellow grass and straggling brown lines of hedged hawthorn and avenues of decrepit elms. The place is Mt Noorat in the Western districts of Victoria, Australia, far from the English landscape that it tries to copy. It fails; this place isn't England.

Halfway up the cinder slope a tuft of trees – gums and she-oaks marks the spot where William Martin has made his home, and where he rejects the English retrospective idioms of Rosemary Verey, Gertrude Jekyll and her latter-day Australian mimic, Edna Walling. He's not concerned with Bali style gardens or Japanese gardens or Renaissance gardens. There are no Romantic, green and lush Summer idylls here. Instead William takes as his starting point for creativity his own knowledge and experience of this place: summer dry, winter wet, yellow grass, green grass, the constant black of rocks and soil; the wild wind that blows incessantly, the huge blue sky and the low dark

clouds; the exposed nature of the hillside and the distant views below of an ancient landscape that hasn't succumbed to the tender nurturing of gardener and grazier. To this understanding he applies his personal vision of what a garden would be here in this place. It is a vision that encompasses a profound interest in making a garden as opposed to simply living in the bush. "Wigandia" is a garden. It moves with the rhythm of the seasons that blow across it. The rhythm is scored by William into a musical episode, an opera buffo perhaps – light, witty, a little risqué, a bit provocative with its own parts in presto – ephemeral winter and spring flowers, forte – the towering furcreas and dasylirions, piano – the feathery grasses, adagio – long drifts of low orange, red, brown, grey succulents, pizzicato – spiky aloes and agaves, alla breve – the short, sudden impact of colletia and carissa, accelerando – high and blowzy buddleja, giant reed and melianthus, crescendo – the overpowering landscape, largo – the slow passage of the sun and shadows across the garden. The movements and entire composition are orchestrated according to William's intimate observation of each botanical and climatic "instrument" and each part played by him with consummate skill. Every piece is just so – balanced but poised to strike off new ideas and fresh associations; invested with sculptures and installations that add spark and sharpness to the scene as progress is made around the whole creation. Each point and counterpoint adding to the mounting thrill of realisation that this place is of this place, like no other – unique and true to itself just like a composition by Philip Glass or a work by Ian Hamilton-Finlay.

项目位置：澳大利亚维多利亚州诺亚特山
景观设计：威廉·马丁
所获奖项："年度最佳花园"等多个奖项

Location: Mount Noorat, Victoria, Australia
Landscape Designer: William Martin
Awards: multiple wins as "Best Garden of the Year"

Europe 欧洲篇

Cany Ash

Principal, MA, Dip Arch, RIBA, FRSA
Ash Sakula Architects (London, England)
www.ashsak.com

"The measure of any great civilisation is in its cities and a measure of a city's greatness is to be found in the quality of it public spaces, its parks and its squares." (John Ruskin, 1819 ~ 1900)

Like other energetic Europeans, we are looking for places that are simply themselves and nowhere else. The Landschaftspark in the Ruhr or the Westergasfabriek in Amsterdam are our places of pilgrimage. Remnants of past human endeavours are more precious to us than new decoration.

Does it make any sense to retain the chimney and shaky walls of a toffee factory burnt out by kids a few years ago on a site for new creative industries? Should we keep the entrance screen to the Victorian cattle hospital along the quay? Does it make sense to celebrate the entrance to a tiny underground train tunnel, which brought coal from the mines on the other side of the city through to the canalside? Crazy though this might seem to the pragmatists, we see ourselves both as custodians of the almost hidden narrative environments on a site, as well as storytellers who are adding another layer to the landscape.

Are we bold enough? Perhaps we are cautious: we have after all during our time as architects experienced the swing of the pendulum from out with the old, to out with the new, to in with the new. And if we are cautious it is because of our attitude towards the past.

Britain has many high quality urban landscapes paid for when it was the world's first wealthy industrial nation, and when pride and patronage in one's town or city was a way to make one's name. Factory owners became town councillors and donated funds for wonderful granite and sandstone pavements and statues. A grand municipal fountain was an important frontispiece to any new town hall, railway station or library. Each proud industrial city had its street lighting, its war memorials and its public gardens and parks.

Since the war these town centres were often deemed overly complacent stage sets for civic life. At that time the modernizers of British cultural life favoured more abstracted work in literature, music and drama, and the architecture to reflect this age was exemplified by the South Bank in London, home to lots of sculptural concrete boxes and walkways. Until about the late 1980s there was a feeling that only certain very special specimens demonstrating the old ways of making cities had to be preserved. Pavements were replaced with cheap concrete, and stone, granite and marble went to landfill or to salvage yards where they could be picked up for a song.

The Victorians had a confidence about place-making rooted in an almost implacable self-belief. Nowadays we like irony, ambiguity, cultural diversity, event-architecture, and find the certainties of old city-making a bit too formal. Still, we like that strong backdrop to play on. Since preservation is less onerous than re creation, those stone pavements are more carefully documented and protected, as if they could lure back an intense street life from the malls.

However the growing realization that quality in the public realm is an essential part of place-branding, not only in European cities but in small market towns, is accompanied by the shock that money will not buy it. A simple square is extremely tricky to conceptualise and realise with conviction. Replica historic street lamps and curlicue benches are now just so passé, seen as pastiche or post-modern junk, which simply tried a little too hard to turn the clock back.

Luckily and interestingly, the best 1960's architecture is now as exotic and precious as the architecture of the Victorians and early twentieth century. Instead of blaming architects for nightmare landscapes in stained concrete, young metropolitans are in love with their patinated surfaces and with the glamour of planar surfaces bathed in sophisticated floodlighting and projection at night. The preservation of many brutalist buildings and landscapes of the sixties is the subject of intense debate. Now that buildings need to be highly insulated we realise it is an era that will not return. Raw concrete is an animal we like more than the tame materials that have replaced it. It spalls and rusts. It is alive. We have been enjoying the plasticity and non-modular character of concrete more and more in our projects: as a new skin to an old tower block in the East End of London; as the inside outside paving to our Carnival Centre; or the implied beach embedding flotsam and jetsam in our Southwold café.

建筑学专业文学硕士，（英）皇家建筑师协会 RIBA、FRSA
Ash Sakula Architects 总负责人（英国伦敦）
www.ashsak.com

任何光辉的文明都寓含于所在的城市，而城市的伟大又都体现在其公共空间、公园和广场的品质上。(John Ruskin，1819 年~ 1900 年)

设计师都在努力打造世界上独一无二的地方。德国鲁尔的 Landschaftspark 和阿姆斯特丹的 Westergasfabriek 正是人们梦寐以求的地方。前人的智慧结晶比设计新作更加令人们驰往。

数年前，几个小孩不小心将一家太妃糖厂付之一炬，保留原址上的烟囱和岌岌可危的墙壁对于创新产业具有什么意义？位于码头旁边通往废弃的维多利亚医院的入口照壁是否应该保留吗？连接城市另一端的矿山与海滨的地下火车隧道，如今是否还需要对其进行修缮？这些对于实用主义者来说似乎很不可思议，但作为景观设计师既要忠实于原有环境，又要通过景观设计来美化环境。

这很胆大妄为吗？其实是十分谨慎的：毕竟当代的设计师正经历着时代的变迁，旧事物不断地被新生事物所替代。这种谨慎源于对过去的态度。

英国处于世界第一工业强国时期，投资建设可以提升捐资人的声望，因此在当时涌现了许多闻名的都市景观。工厂老板当选为镇议员后出资采购上等的花岗岩和砂岩来铺路，并建设雕塑。在市政大厅、火车站或者图书馆前都会修一座漂亮的喷泉作为门面。当时每个知名的工业城市都设有路灯，建有抗战纪念碑和公园。

战争过后，城镇中心被看做安居乐业的港湾。在那个年代，英国现代主义者们更喜欢通过抽象的文学作品、音乐和戏剧的形式来展现他们的生活。利用建筑来反映时代特点的典型代表是伦敦的南区银行，这里汇聚了大量的雕塑和人行步道。直到 20 世纪 80 年代末期，人们才意识到只有少数能体现过去城市建设方式的景观才值得保留。原来的人行道改用便宜的混凝土重新铺设，而被取代的岩石、花岗岩和大理石都被送到废物回收站，再以低廉的价格被回收利用。

在维多利亚时期人们从内心深处对营造场地的能力十分自信。如今，设计师们推崇反讽、含糊、提倡文化多元化，并修建会展建筑，认为老式城区建设方式有些死板。然而，他们也喜欢在原有的景观之上进行修缮，因为这比彻底重建的工作量小，因此这些石路被完好地保存下来，以期能够带动商业街的发展。

人们越来越清晰地意识到提升公共区域的建设对于欧洲的大都市和小镇都很重要，它影响着城市的形象，然而这并不是足够的资金投入就可以做到的。即使一个普通的广场设计也需要技巧，单纯模仿过去的街灯和雕花长椅的设计如今看来是非常落伍的，这种照搬过去的设计风格简直就是山寨，甚至是后现代废品。

有趣的是，20 世纪 60 年代的杰出建筑与维多利亚时期和 20 世纪初期的建筑一样备受珍视。都市里的年轻人不但不反对建筑师给建筑着上怪异的颜色，并且喜爱建筑锈迹斑斑的感觉，青睐夜晚强烈的灯光打在建筑物上光彩夺目的效果，很多野兽派建筑和 20 世纪 60 年代的景观是否需要保留引起了激烈的讨论。建筑是独立于时代而存在的，而我们却无法回到过去。现在很多建筑都经过精雕细琢，但设计师们还是偏爱那些用未经加工的材料堆砌的粗糙的建筑，虽然它非常容易脱落和生锈，但它是有生命力的；由于它可塑性强且不拘一格，设计师们越来越喜欢在工程中使用，如伦敦东部一座古老大楼的新外立面，狂欢中心室内和室外的地面铺设，索思沃尔德餐厅一块点缀着海上漂浮物的人工沙滩。

围墙之内有洞天——马克思·霍尔植物园的围墙花园

A Big World on the Other Side of the Wall—The Walled Garden at Marks Hall Arboretum

翻译　董桂宏

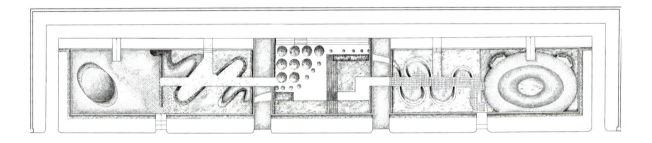

马克思·霍尔庄园的占地面积约为 81 万平方米，包括历史悠久的鹿园、17 世纪挖掘的人工湖和 18 世纪建成的围墙花园，至今仍保留着英国古老的撒克逊风格。庄园的建筑早已化为历史尘埃，这片场地也在 1971 年被改造成为一座植物园并对公众开放。然而马克斯·霍尔庄园的围墙花园因无人料理而杂草丛生，设计师受命将这里重新打造成一个令游客流连忘返的大"蜜罐"。

该项目中的围墙均是 18 世纪遗留下来的古墙，设计师新建了一系列的花室，花室的南面紧挨着人工湖。从人工湖向对面望去，游客的视线会被一道与古墙平行的角树篱挡住，使得这些新建的花室在角树篱后若隐若现。碧绿的角树篱与古墙橙黄的粘土砖在色彩上形成了对比。角树篱掩映着花园中繁茂的植物和新旧建筑，增加了花园的神秘感与庄严感，

游客只有从入口进入花园时才能看到隐藏着的围墙花园。

青绿色的欧洲赤松华盖如伞，荫蔽着古墙。绿色的欧洲赤松、橙黄的古墙以及浅绿色的角树篱组成三条色彩带，集中展现了埃塞克斯郡的平原风光。

该项目由彼此衔接的三部分组成，西侧的小花园中有一座泥塑；第二座小花园叫做斯福瑞斯花园（Garden of Spheres），其正中是一座石雕，与西侧小花园的泥塑相呼应，花园四周的篱笆高低起伏。在花园的阶地上可以眺望美丽的湖景，此处视野开阔，与封闭的花室形成对比。第三座小花园中建有一座矮墙，矮墙建在高高的角树篱之中，矮墙的尽头是一间花室，周围是下沉的池塘，形成高低起伏的空间布局，与第二座小花园中起伏的篱笆交相呼应。

Marks Hall's 200-acre estate has a long history that encompasses a Saxon manor and a deer park long before the 17th century lake and 18th century walled garden were built. The house no longer exists but since 1971 the grounds have been developed into an arboretum and are open to the public. Schoenaich Landscape Architects were commissioned to design a "honeypot" and a point of destination within the old walled garden, which was no longer used for growing crops.

A series of garden rooms are set within the 18th century walls, which are bordered by a lake on its southern side. The new gardens are hidden from views across the lake by a hornbeam hedge running parallel to the wall. The hedge creates a green line set against the rich orange colour of the old clay bricks. It hides the intricate gardens and retains the dignity of the space. The hidden gardens are only revealed as one enters through the gate.

The crowns of blue green Scots pine tower above the wall. The three bands (dark pines, orange wall, pale green hedge) reflect the linearity of the plains of Essex.

The gardens are linked by a metaphorical movement, which is "triggered off" in the first western garden by an earth sculpture. In the second garden a hedge meanders though the space, turning into stone in the Garden of Spheres in the centre of the intervention site. In contrast to the enclosed garden rooms, this terrace opens up the view across the lake.

The meandering theme is continued with a low wall in the next garden, culminating in an inward looking garden room, set within tall hedges around a sunken pool.

项目位置：英国埃塞克斯

客　　户：托马斯·菲利普·普埃斯信托公司

总 成 本：185 000 英镑

占地面积：500m²

项目日期：2000 年～ 2003 年（项目完工后定期维护）

景观设计：Schoenaich 景观设计事务所

Location: Essex, England

Client: Thomas Phillips Price Trust

Budget: £185,000

Surface Area: 500 sqm

Project Dates: 2000~2003 (seasonal updating thereafter)

Landscape Design: Schoenaich Landscape Architects

回归自然——伦敦伯德菲尔德公园

Back to Nature—Potters Field Park in London

翻译 刘建明

在重塑伦敦与泰晤士河和谐关系的系列景观项目中，伯德菲尔德公园无疑是极为重要的城市景观元素之一。伯德菲尔德公园将这座城市非比寻常的历史融入到这个备受人们喜爱、极度细化的城市空间。

20世纪的设计理念由标新立异到返璞归真，景观设计也趋于回归自然。最近，花园作为缩微自然的密集空间、度过闲暇时光的天堂和满足人们亲近自然愿望的空间又重新回归，并成为城市象征的表现手段。

伦敦向来就有建造规模超大的公园的传统，因此新建的城市空间的确没有什么值得炫耀。政治与文化气候的变迁，给这座城市的复兴烙上了清晰的印记。在经过多年来的忽视与不合理利用后，借助一系列的文化建筑与公共空间，泰晤士河再度成为了伦敦的生命线。该项目就是这些新开发项目中的一个，其所占据的独一无二的河岸位置使其拥有极为优越的视角，将伦敦一些标志性的历史名胜尽收眼底——塔桥、伦敦塔以及千变万化的伦敦城市轮廓线。

该设计方案朝向居民区，就像是一处邻家后花园，沿着阶梯就可以一步一步地接近泰晤士河。巨大的工字梁结构支撑着一条通向花园的永久性通道；沿着倾斜的河道种植适于举行各种盛会的草坪。设计师在入口处设计的两扇平行的装饰性铸铁大门和一路延伸的树篱更加突出了主通道，并且在花园和道路交叉口之间设计了一个安全的缓冲带。铸铁大门的图案源自该地区作为英国代夫特陶器产地的悠久历史。类似的具有象征性历史意义的景观还表现在该项目的多处细节中，例如主对角线通道旁的大型坐椅也是按照英国代夫特陶器的图案来排列的。

由荷兰设计师 Piet Oudolf 担纲设计的多样的草本植物在公园中随处可见。枝繁叶茂的白桦林柔化了伦敦市政府的标志性建筑——形似穿山甲的办公大楼所带来的视觉冲击；盛开的樱桃树遍布广场周围，与塔桥浑然一体。

该项目最具挑战性的设计难点是如何在相对狭小、过度利用的空间里调和不同群体之间的冲突，例如居民、工人、访客和游客。为了满足各类群体的需求，公园24小时全天候开放，所有的主路都有路灯在夜间照明。公园边缘和"摩尔伦敦"开发项目之间有一个小亭子，可为人们提供休憩之所。

The Potters Field Park inserts a vital urban element into a series of projects intended to re-vitalize London's relationship to the River Thames. The park integrates the site's unique history into a richly detailed space that caters to both the public and private spheres.

During the 20th century the particular became generic, landscape became environment and a delicate green tissue turned into a monstrous green blanket. Recently the garden as urban typology has been re-discovered; the garden as intensified space of condensed nature, safe heavens for idle strollers and voluptuous voids of hidden desire.

Whilst London has a legacy of oversized and understated public parks, it has little to show for new urban spaces. Changes in political and cultural climates, however, have resulted in a remarkable urban renaissance. After years of neglect and visual abuse, the River Thames once again becomes London's

lifeline, with a string of cultural buildings and civic spaces. Potters Field Park is part of this series of new developments. Its unique riverfront location provides open views towards some of London's most iconic historic monuments such as Tower Bridge, the Tower of London, and the ever-changing City of London skyline.

The design is an intimate neighbourhood park that faces the residential areas and gradually opens up towards the River Thames with a series of stepped terraces. The lawns, suitable for various events, descend down towards the river walkway. A large I-beam structure supports a new permanent gateway for the park. Combining two parallel ornamental cast iron gates and an extended hedge, the entrance reinforces the main path, and allows a safe transition between park and road crossing. The cast iron gates have a pattern derived from the site's history as an English Delftware production area. Similar emblematic historical references appear in various details throughout the park. For instance, a large seat with English Delftware pattern aligns the main diagonal pathway.

A spectacular variety of herbaceous plants and grasses designed by Piet Oudolf of the Netherlands appear throughout the park. The visual impact of the Greater London Authority's building, an iconic structure also known as "armadillo", is softened by a veil of multi-stemmed birch trees. Cherry blossom trees fill a square, and create an intimate transition with Tower Bridge.

One of Potters Field Park's most challenging regenerative aspects will be the reconciliation of different user groups such as residents, workers, visitors, and tourists in a relatively small and intensively used space. To account for the various groups, the park is open 24 hours, and all main paths ways are lit at night. A small kiosk between the park edge and the More London development provides facilities for both park users and office workers.

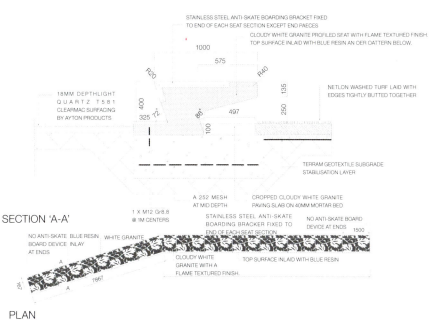

STAINLESS STEEL ANTI-SKATE BOARDING BRACKET FIXED
TO END OF EACH SEAT SECTION EXCEPT END PAECES

CLOUDY WHITE GRANITE PROFILED SEAT WITH FLAME TEXTURED FINISH.
TOP SURFACE INLAID WITH BLUE RESIN AN OER OATTERN BELOW.

1000
575

R20   R40

18MM DEPTHLIGHT
QUARTZ T581
CLEARMAC SURFACING
BY AYTON PRODUCTS

400   135

NETLON WASHED TURF LAID WITH
EDGES TIGHTLY BUTTED TOGETHER

325 72°   86°   497   250

100

TERRAM GEOTEXTILE SUBGRADE
STABILISATION LAYER

A 252 MESH
AT MID DEPTH

CROPPED CLOUDY WHITE GRANITE
PAVING SLAB ON 40MM MORTAR BED

SECTION 'A-A'

1 X M12 Gr8.8
@ 1M CENTERS

STAINLESS STEEL ANTI-SKATE
BOARDING BRACKER FIXED TO
END OF EACH SEAT SECTION

NO ANTI-SKATE BOARD
DEVICE AT ENDS

1500

NO ANTI-SKATE BLUE RESIN
BOARD DEVICE INLAY
AT ENDS

WHITE GRANITE

CLOUDY WHITE
GRANITE WITH A
FLAME TEXTURED FINISH.

TOP SURFACE INLAID WITH BLUE RESIN

7867

PLAN

项目位置：英国伦敦	Location: London, England
客　　户：More London Ltd., 伦敦池的合作伙伴—— 南华克市政府	Client: More London Ltd., Pool of London partnership, Southwark Council
预　　算：460 万美圆（350 万英镑）	Budget: US$4.6 million; (€ 3.5 million)
占地面积：17 000m²	Area: 1.7 ha.
设计时间：2003 年 ~2007 年	Project Design: 2003~2007
施工时间：2007 年 ~2008 年	Project Implementation: 2007~2008
景观设计：GROSS.MAX Landscape Architects & Piet Oudolf	Landscape Architecture: GROSS.MAX Landscape Architects In collaboration with Piet Oudolf

棕地改造——毕尔巴鄂河畔

Brownfield Regeneration—Bilbao Waterfront

翻译　张晶　董桂宏

西班牙毕尔巴鄂市的阿班多尔巴拉区 (Abandoibarra) 总体规划由 Bamori 设计师事务所 (Bamori Associates) 担纲设计。这一改造工程为古老的工业港口城市带来了新的生机。Bamori 设计师事务所同佩里·克拉克·佩里建筑师事务所 (Pelli Clarke Pelli Architects) 组成了专业设计团队,承揽公园规划以及所有开放空间、街道、人行道和广场的设计。设计师本着保护生态环境的设计理念,特别注重扩大绿色空间,为此他们将原来的一条高速公路改造为穿插着多条人行横道的林阴大道。一条便捷的轻轨连接两大文化中心——建筑大师弗兰克·盖瑞 (Frank Gehry) 设计的古根海姆美术馆 (Guggenheim Museum) 和歌剧院。铁轨从大片的草坪中横穿而过,保证了绿色空间的延展性。

新建的线性河畔公园与建于 19 世纪的多娜·卡西尔达公园 (Doña Casilda) 合二为一,城市中心这块 50 万平方米的土地成功地被改造为绿色休闲空间。考虑到河水的潮涨潮落,同时也为了增加行人的活动空间,滨河路特意设计了两个不同的水平面:水漫区和无水区。河水的水位低时,水漫区堤岸上的台阶就完全显露出来,行人可以在此休憩。2005 年,总体规

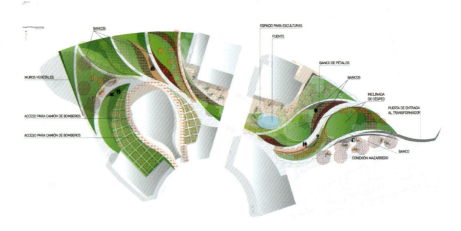

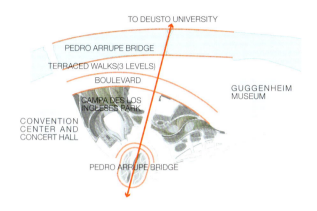

TO DEUSTO UNIVERSITY

PEDRO ARRUPE BRIDGE

TERRACED WALKS(3 LEVELS)

BOULEVARD

CAMPA DES LOS INGLESES PARK

GUGGENHEIM MUSEUM

CONVENTION CENTER AND CONCERT HALL

PEDRO ARRUPE BRIDGE

划中的里维拉公园 (the Parque de la Rivera) 设计方案赢得了水城威尼斯建筑双年展 (Biennale di Venezia) 最佳工程奖的殊荣。

到目前为止, Bamori 设计师事务所还承揽了毕尔巴鄂河畔另外两项工程的设计工作:

Euskadi 广场 (Plaza Euskadi) 的设计将城市各种不同风格的建筑有机地统一起来, 仍然延续了绿色空间的设计理念。这里有着不同时期的城市开发遗迹: 几幢历史悠久的居民楼, 一个空间架构的天篷, 一处地铁站, 酒吧和餐馆林立的林阴路。这些截然不同的元素被出色地融合到一起, 打造了依地势、植被和建筑物而建的多样化的公园空间。

Campa de los Ingleses 公园建于毕尔巴鄂市的古根海姆美术馆 (Guggenheim Museum) 附近, 是连接阿班多尔巴拉区 (Abandoibarra) 与内维隆河 (Nervión River) 的纽带。公园两侧是蜿蜒起伏的小路和台阶。地势的起伏使公园内部形成 10m 的高差。台阶、斜坡、楼梯和围墙层层叠叠, 彼此相连, 将马萨雷多大街 (Mazarredo)、德乌斯托大桥 (Deusto Bridge)、巴斯克广场同周围的建筑物以及内维隆河衔接得天衣无缝, 构成了毕尔巴鄂独特的城市一景。

Balmori Associates' Abandoibarra Master Plan for the post industrial port city of Bilbao, Spain weaves new development into the old city. Teamed with the office of Pelli Clark Pelli Architects, Balmori Associates created park guidelines and designed all open space, streets, sidewalks and plazas. Placing particular emphasis on expanding the amount of green space and incorporating sustainable design practices, a high-speed roadway was "slowed down" and turned into a boulevard with multiple pedestrian crossings. A light rail now connect the two main cultural centers of the development: Frank Gehry's Guggenheim Museum and the opera house. Running on wide swaths of green lawn, this rail line gives continuity to the green space.

A new linear park links the 19th century Doña Casilda Park with the river's edge and in the process effectively turns the entire 50 acre section of the city into an expanded green and recreational space. To accommodate the river's natural cycles and provide additional pedestrian space, the edge of the river has two distinct levels, a floodable and non-floodable walkway. On the floodable level, when the tides are low, steps emerge from the bulkhead

allowing pedestrians to sit down by the water. In 2005, this piece of the master plan, the Parque de la Rivera, received the Special Award "Citta d'Aqua" of the Biennale di Venizia for the Best Project.

Balmori has since been designing two more projects in the waterfront :

Plaza Euskadi emerges as a continuous green gesture that unifies the diverse built elements of this urban site. The site is fragmented by remnants of various stages of urban development: historic residential buildings, a space frame canopy, an underground subway station, and a tree-lined street with bars and restaurants. These disparate pieces have been incorporated into a variety of park spaces through topography, planting and built elements.

Campa de los Ingleses Park flows from the Guggenheim Bilbao Museum, unifying the Abandoibarra area of Bilbao and the Nervión River. The park's defining lines mark undulating paths that pull up to create a series of curving terraces. These topographic waves mediate a 10m elevation difference across the park. The terraces, ramps, stairs and walls flow into one another to sculpt a park that gracefully integrates the Mazarredo, Deusto Bridge, and the Plaza Euskadi with surrounding buildings and most importantly the Nervión River into a seamless urban experience.

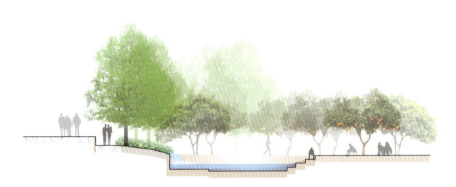

项目位置：西班牙毕尔巴鄂
客　户：Sociedad Bilbao Ria 2000

Euskadi 广场：
占地面积：10 000m²
设计团队：Balmori Associates，Cao | Perrot Studio
景观设计：Balmori Associates

Abandoibarra 总体规划：
占地面积：300 000m²
设计团队：Balmori Associates，Pelli Clarke Pelli Architects，
　　　　　Aguinaga & Associates Architects
景观设计：Balmori Associates

Campo de los Ingleses 公园：
占地面积：25 000m²
设计团队：Balmori Associates
景观设计：Balmori Associates，RTN Architects，Lantec Engineers

Location: Bilbao, Spain
Client: Sociedad Bilbao Ria 2000

Plaza Euskadi:
Site Size: 10,000m²
Design team: Balmori Associates and Cao | Perrot Studio
Landscape Design: Balmori Associates

Abandoibarra Masterplan:
Site Size: 300,000m²
Design team: Balmori Associates, Pelli Clarke Pelli
　　　　　　　Architects, Aguinaga & Associates Architects
Landscape Design: Balmori Associates

Campo de los Ingleses Park:
Site Size: 25,000m²
Design team: Balmori Associates
Landscape Design: Balmori Associates, RTN Architects,
　　　　　　　Lantec Engineers

一地两用——市场与停车场的转换

Dual Function—The Exchange of Market and Parking

翻译　董桂宏

克佩尼克市历史悠久，其中心城区亟待改造。克佩尼克市是一个典型的中世纪风格城市，以其历史悠久的建筑而闻名，但城市中的棕地和待开发场地也是该市的一大特征。虽然克佩尼克市的内城街道和广场经过改造焕然一新，但是城市的结构仍不完善，未来如何发展仍是一个未知数。KAiAK艺术项目的宗旨是根据短期需要设计一些暂时性的项目，以加强城市建设的创新动力，而这些暂时性的项目可以在适当的时机被改造成实体建筑。基于这一宗旨，设计师采取了一系列目标一致的设计干预。

场地中两条街的交叉口处被用做非正式的停车场。该项目考虑到城市发展和村镇发展互相补充的需要，场地主要被用做农贸市场，其他活动也可以在此举行。原有的柏油马路被改造成现代感十足的红色路面，既能用来停车，也能在此搭建售货棚。地面的红色不是一成不变而是根据色彩的光谱形成颜色上的渐变。路面上标示出的微微弯曲的数字网格，给场地增添了一种与众不同的动感。

该场地具有双重功能，既可以停车又可以作为临时市场，而这两种功能之间的转换方式简洁直观：当场地作为临时市场使用时，一个特大号的浅红色遮阳伞伫立在场地上作为售货亭，遮阳伞的颜色与新鲜柿子椒的颜色一模一样，娇艳欲滴；而当场地作为停车场使用时，人们就将遮阳伞折起收好，这代表着此时可以停车。

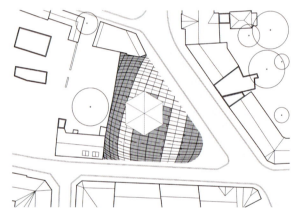

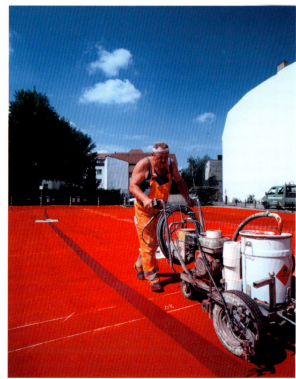

The historic centre of the city of Köpenick is subject to change. At core of medieval structure, Köpenick is characterized by its historical buildings, but also by brown-fields and empty lots. While the inner city's streets and squares have respectfully been renovated, the gaps in the urban texture keep asking questions about the future. The aim of the KAiAK art-project was to search for temporary uses which may be replaced by a building at some point, but generate a creative impulse for Köpenick's urban development. A network of design-interventions was to support its re-conception.

The project-site at the corner between two streets was informally used as a parking lot. Introducing a higher degree of complementarity and urbanity, the site is now to be used alternately for a farmer's market, or other activities. The existent asphalt surface was turned into a red urban parquet floor on which cars and market-stalls take turns in presenting themselves. Laid out in a differentiated spectrum of reds, a fine, slightly warped grid of lines and ciphers adds an unusual and invigorating dynamic to these uses. The exchange of functions is being signalled with a gigantic gesture. On market days, an oversized sunshade, lacquered in waxy red, reminiscent of a fresh bell-pepper, turns into a market pavilion. When the site is used as a parking lot, it is folded back up again.

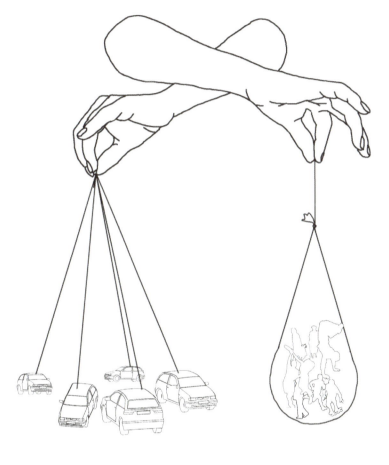

项目位置：德国柏林克佩尼克
客　　户：Stadtkunstprojekte e.v.
占地面积：1022m²
项目日期：2006 年 ~ 2007 年
景观建筑与景观设计：Topotek 1
所获奖项：克佩尼克艺术与建筑设计奖第一名

Location: Berlin-Köpenick, Germany

Client: Stadtkunstprojekte e.v.

Site Size: 1,022 m²

Project Dates: 2006~2007

Landscape Architects/Artists: Topotek 1

Awards: 1st Prize Kunst und Architektur in Alt-Köpenick

(KAiAK) (Art and Architecture in Alt-Köpenick)

生态公园
Ecological Park

翻译 董桂宏

　　该项目场地原本是一块退化了的湿地，Groupe Signes 提出在马恩河 (Marne River) 的臂弯建立一座生态公园，其宗旨是保护马恩河流域的动植物群落和生物多样性。该项目占地面积约 70 万平方米，通过观景屋与观景平台拉近人与自然的距离。

　　该项目靠近马恩河的河汊口，设计师根据客户的需要将场地的各种元素重新组合。天然的河岸生态系统和考古发现的史前遗留为建设河流公园提供了有利的条件，并且以一种和谐的方式展示出人类与河流的关系。

　　设计师将不同的水生生态系统有机地结合起来，如将 U 形水体生态系统（由于宽阔的河流主水体蜿蜒流动而形成的 U 形水体）、池塘生态系统和冲击区生态系统等多个水生生态系统有机结合成复杂的大型水生系统。该项目采用了全新的绿化技术和绿化管理方式，使其成为一座研究生态与可持续发展的大型实验室。桥渡、高于地面的河岸木栈道和木质观景台都使游人能够与马恩河亲密接触，以及近距离地观察飞经此地的候鸟。

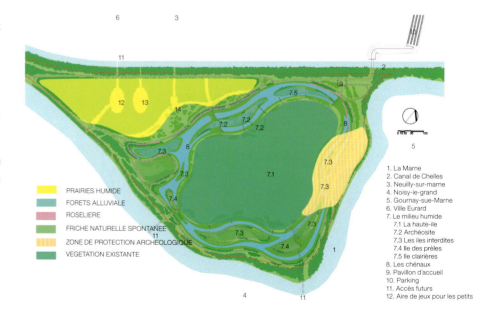

PRAIRIES HUMIDE
FORETS ALLUVIALE
ROSELIERE
FRICHE NATURELLE SPONTANEE
ZONE DE PROTECTION ARCHEOLOGIQUE
VEGETATION EXISTANTE

1. La Marne
2. Canal de Chelles
3. Neuilly-sur-marne
4. Noisy-ie-grand
5. Gournay-sue-Marne
6. Ville Eurard
7. Le milieu humide
　7.1 La haute-ile
　7.2 Archéosite
　7.3 Les iles interdites
　7.4 Ile des prêles
　7.5 Ile clairière
8. Les chénaux
9. Pavillon d'accueil
10. Parking
11. Accès futurs
12. Aire de jeux pour les petits

Groupe Signes proposes an ecological park on a former arm of the river Marne, in which a degraded wetland was rehabilitated to strengthened biodiversity of vegetation and fauna. This ecological park with approximately 70 hectares of property welcomes the public through a network of watch houses and observation decks.

Situated near a fork in the river, this section of the river was reorganized according to the department's priorities. The observation of the natural dynamics of ripicole ecosystems, and the discovery of pre-historic vestiges on site have created new opportunities to project the river park and notably to recount the relationship between Man and its rivers.

Groupe Signes' intervention puts in place different aquatic ecosystems such as oxbow lakes (a U-shaped body of water formed when a wide meander from the mainstem of a river is cut off to create a lake), ponds and alluvium basins, among others, articulating all these into larger, more complex systems. New greening techniques and management supported by the natural dynamics of the site were tried out and applied, transforming the Parc de la Haute-Ile in a research laboratory on issues of ecology and sustainable development. River crossings, suspended boardwalks and observation decks allow visitors to be in close contact with the park's mutations, and with passing migratory birds.

项目位置：法国 Neuilly—Sur—Marne
客　　户：Département de Seine Saint—Denis
成　　本：1520 万欧圆
占地面积：70 万平方米
项目时间：1998 年～ 2008 年
景观设计：SIGNES Paysages 与景观设计师
　　　　　Allain Provost

Location: Neuilly-Sur-Marne, France
Client: Département de Seine Saint-Denis
Budget: €15,2 million
Surface Area: 70 ha.
Project Dates: 1998~2008
Landscape Architects: SIGNES Paysages with landscape
designer Allain Provost

"光之花园"——阿尔巴尼庭院

"Garden of Light"—The Albany Courtyard

翻译　董桂宏

"进行设计时，我们十分尊重场地的遗产——这一设计也体现了我们对历史遗产的关注。新设计与旧特色可以和谐地融为一体。我们希望保留阿尔巴尼原有的特色，创造出独具魅力的新设计，既能够表达对历史的尊重，又能够与时俱进，创造出具有现代感的设计。"

——安迪·托马斯（BCA 景观）

阿尔巴尼庭院位于英国利物浦市一处受到文化保护的建筑之中，庭院原已荒弃，经过精心设计之后，如今已形成一座"光之花园"。在设计之前，设计师们曾分析过多种多样的设想，包括适宜居住的豆荚造型等，但最终选定一系列现代感极强的涡卷形和螺旋形的造型。该设计源于阿尔巴尼公司的建筑师 JK 柯灵的设计构想，JK 柯灵对中世纪的叶子和花卉造型情有独钟，因此，叶子与花卉造型成为石质雕刻作品和石膏栅栏

的基本图案，其艺术表现形式多种多样。庭院内全新的坐椅、灯饰和格架都是特别订制的。为使设计达到理想的视觉效果，设计师们征求了欧洲各国的家具专家和灯饰专家的意见。

盘旋上升的枝形吊灯

每一盏枝形吊灯的螺旋形结构都由 2250 个 14mm 的施华洛世奇铅质玻璃水晶珠子组成。枝形吊灯的直径为 1m，高约 1m，重 25kg。铅质玻璃水晶是质量上乘的水晶材料，常用于制作豪华精致的珠宝。在悬线的作用下，枝形吊灯仿佛悬浮在场地上空一般。基于环保考虑，不锈钢悬线的直径均不超过 4mm。

涡卷形坐椅

经过雕刻的曲线形坐椅由浇铸的红色玻璃钢建造而成，

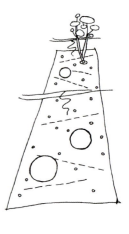

位于枝形吊灯的正下方。夜晚，闪亮的 LED 地下照明使坐椅看起来仿佛是盘踞在地面的"Yorkstone"标记上。LED 地下照明的电源和坐椅的地面安装细节都隐藏在铝质坐椅腿之中。坐椅腿的直径各不相同，根据椅面的大小按比例改变。每一个坐椅都由五个部分构成，以便将它分成五部分从前门搬进庭院内。

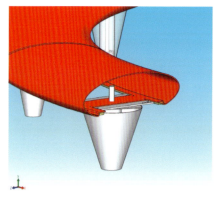 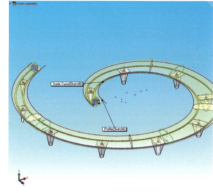

圆锥形格架

庭院内所有设施都根据庭院原有的特色进行设计。该项目需要在场地中设置树池，但设计师们并没有破坏场地原有的铺装，而是将原有的煤槽改造成了树池。设计师们还在树池的上方安装了 16 个 3m 高的圆锥形格架，种植在树池中的常春藤会沿着螺旋形的格架向上生长。圆锥形格架由可以发光的纤维光学材料构成，发出闪闪的光芒，与坐椅和枝形吊灯相互辉映，共同形成令人叹为观止的庭院夜景。庭院内所有的灯光均由设置在保安室中的调光系统进行控制。

修复工程

走进庭院，首先映入眼帘的是场地原有的铸铁人行桥，它将庭院一分为二。螺旋形的铸铁阶梯连接着铸铁人行桥，其下方的照明装置发出微光，增加了阶梯的亮度。游人接着会看到下沉空间中的枝形吊灯和涡卷形坐椅，这些新的设计元素与场地原有的特色相得益彰。设计团队与利物浦市的遗产保护专家密切合作，确保庭院原有设施的修复工程顺利进行，并且确保在庭院外面看不到铸铁人行桥。场地原有的铸铁栏杆经过精心修复，既增加了场地的现代感又确保了游人的安全。

"We have always been very aware of the great heritage that we were working with here and hopefully we have shown that with care and attention to detail – new and old can exist together in harmony. Ultimately we wanted to add to the Albany's uniqueness and create something magical that celebrates the past, but looks to the future"

– Andy Thomson

Rescued from dereliction, the courtyard within this protected building in Liverpool (UK) has now become the setting for the ground breaking "Garden of Light". The design team explored a variety of concepts, including habitable pods within the court, before eventually settling on a design based around a series of scrolling and spiralling contemporary forms. These were inspired by the Albany's original architect JK Colling and his passion for medieval foliage and flower illustrations, which can be seen in various motifs within the building's carved stonework and cast railings. The new seating, lights and trellis structures are all bespoke and unique to the Albany. In order to make the cutting edge design a reality and realise their vision, the designers enlisted the services of specialist furniture and lighting manufacturers from across Europe.

Spiral Chandeliers

Every chandelier comprises 2250 Swarovski Strass 14mm crystal beads set on a chromed helix. Each is approximately 1 metre tall by 1 metre in diameter, and weighs 25 kilos. Swarovski Strass crystal is considered to be amongst the finest quality crystal in the world, normally used for exquisite and luxurious jewellery. The chandeliers seem to float above the courtyard on a catenary's wire system. To satisfy conservation concerns none of the stainless steel hanging wires could be above 4mm in diameter.

Scroll Seating

Sinuous and sculptural seating constructed from moulded red

GRP is situated directly beneath each chandelier at ground level. At night dramatic LED under-lighting makes the seats appear to hover above the original Yorkstone flags. The power for the lights and ground fixing details are all hidden within the spun aluminium legs. All the legs have varying diameters – scaling down as the section of the seat decreases. Each seat was constructed in 5 parts, so it could be carried through the front door and down a flight of steps in to the basement court.

Trellis Cones

All the new elements within the court tread lightly on and around the original features. Rather than break out the courtyard paving for new tree-pits - the existing coal holes in the ground are utilized as planters - above which are fixed sixteen, 3 metre-tall trellis cones. As they grow, the ivies are being trained around the spiral. Each is illuminated with "shimmer" fibre optics, which glimmer to accompany the seating and chandeliers. All the lights in the court can be controlled by a dimmer system within the concierge. Night-time viewing is essential.

Restoration

On entering the courtyard, the visitor is met by the sight of the fully restored original cast-iron bridge that bisects the courtyard. The cast-iron spiral staircase that leads to the bridge is also illuminated with subtle uplighting. The chandeliers and seating are revealed upon descending into the space - the new elements harmonizing with the original features. The team worked closely with Heritage and Conservation experts within the city to ensure the careful restoration of original features and the retention of the view of the cast iron pedestrian bridge from the street. Within this context – elements such as the cast iron balustrade were refined to ensure compliance with modern safety regulations.

项目位置：英国利物浦默西塞德郡

客　　户：英国阿尔巴尼资产有限公司（阿尔巴尼建筑）

总 成 本：22 万美圆（15 万英镑）

占地面积：500m²

项目日期：2005 年～ 2006 年

景观设计：BCA 景观

灯光设计：Crucible ID（牛津大学）、Igguzini（意大利）、
　　　　　Light Lab（伦敦）

所获奖项：2006 年英国国家玫瑰设计奖最佳场地规划奖
　　　　　2006 年英国国家玫瑰设计奖最低成本项目
　　　　　2006 年 4 月最佳居住区发展奖（英国每日邮报
　　　　　和 Echo Property 联合评选的奖项）

Location: Liverpool, Merseyside, England

Client: Albany Assets Ltd, UK (The Albany Building)

Budget: US$220,000 (£150,000)

Surface Area: 500 sqm

Project Dates: 2005~2006

Landscape Design: BCA Landscape

Lighting Design: Crucible ID (Oxford), Igguzini (Italy) and Light Lab (London)

Awards: Best Place Making, UK National Roses Design Awards 2006

Best Low Cost Project, UK National Roses Design Awards 2006

Best Residential Development April 2006 (The Daily Post and Echo Property Awards, UK)

绿色公园——蒙德哥公共绿地

Green Park—Public Green Area of Mondego

翻译　李沐菲

根据几项细部设计，该场地的功能性得到了提升。这些设计形成了一个系统，即在科英布拉城的城市河道公园和蒙德哥河两侧的行人通道系统。该项目达到这种干预的主要目的有两个：修建堤坝使修道院地区免受洪水的侵袭，并创建一个公共绿地，引导位于城市左岸的人们由地下通道来到蒙德哥绿色公园，因为路面上的交通很拥挤。

该项目设计了三种路径：一种路径可以直接进入蒙德哥绿色公园；另一种路径是穿过干预较低的区域，即保护 Sta. Clara 修道院新建的堤坝；最后一种路径则延伸到堤坝的顶部，在那儿可以观赏到修道院的美丽景色，还可以通向纪念碑入口处的新巴士停车场。

由于修道院区和人行通道之间的水平高度不同，产生了几个斜坡，其陡峭程度不同，但方向都朝阳。这些斜坡的主要植被覆盖物是草地，草地的边缘是乔木和灌木丛，需要在那里设置一些限制行人穿越的措施；树木则减轻了附近交通对该公共空间的干扰。

一个宽阔的石板铺装区域能使人们通向这几条人行道，也可以用做坐位区。石板可以用在人行道上或是连接相邻区域，除了石板路面，还有木板路以及由水泥和草地混合铺设的路径。

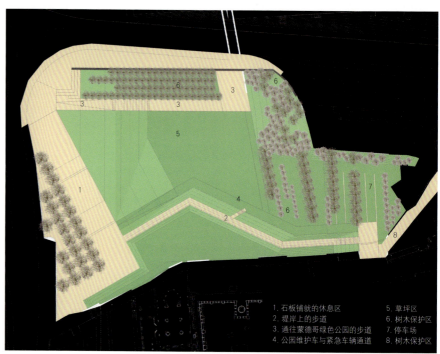

1. 石板铺就的休息区　　5. 草坪区
2. 堤岸上的步道　　　　6. 树木保护区
3. 通往蒙德哥绿色公园的步道　7. 停车场
4. 公园维护车与紧急车辆通道　8. 树木保护区

The intervention lies within a territory that has been re-qualified, in accordance to several Detail plans. These plans promote a system of urban river parks and pedestrian pathways on and between both sides of the Mondego River, in the city of Coimbra. There are two main objectives to achieve with this intervention, to protect the convent area from river flooding, building an embankment/dike, and to create a public green area that will lead people from the left bank of the city to the Mondego's Green Park, through a tunnel underneath an existing barrier, a road with a rather intense traffic.

In this new public space there are three types of pathways: one that allows the access to the Mondego's Green Park, another that crosses through the lower area of the intervention, near the new built embankment/dike that protects Sta. Clara's Convent; and the latter, one, that stretches on top of the embankment/dike, and therefore enables a beautiful view of the referred Convent. This latter pathway connects the new bus parking area with the entrance of the monument.

As a result of the different spot height levels between the Convent area and the pedestrian pathways, several slopes with different degrees of steepness and solar orientations are generated. The vegetation cover of these slopes is mainly grassed, eventually with trees and shrubs where crossing over prevention is needed. Trees are also used to protect this new public space from the adjacent traffic circulation.

A large, flagstone paved, area enables the access to the several pathways and may also be used as a seating area. The flagstones are used in some pathways and in the areas that connect them. Besides the flagstone pavement, there are wooden pathways and concrete grass pavers.

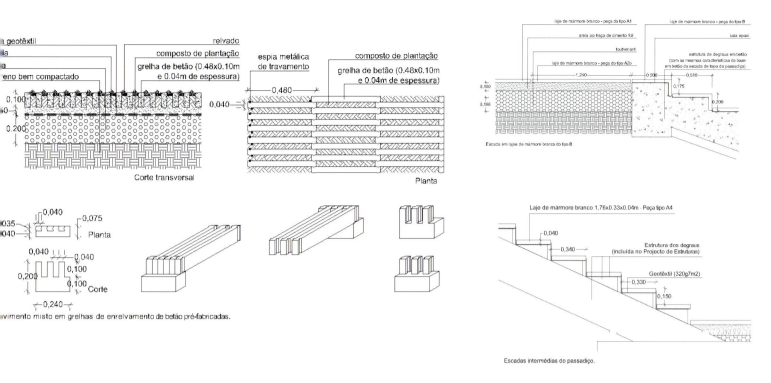

laje de mármore branco - peça do tipo A1
laje de mármore branco - peça do tipo B
areia ao traço de cimento 1:8
cola epoxi
toutvenant
estrutura de degraus em betão
(com as mesmas características da base
em betão da escada de topo do passadiço)
laje de mármore branco - peça do tipo A2b

a geotêxtil
relvado
ia
composto de plantação
eno bem compactado
grelha de betão (0.48x0.10m
e 0.04m de espessura)
espia metálica
de travamento
composto de plantação
grelha de betão (0.48x0.10m
e 0.04m de espessura)

0,100
0,200
0,480
0,040
0,100
0,168
1,240
0,500
0,510
0,175
0,200

Corte transversal
Planta
Escada em lajes de mármore branco do tipo B

035
040
0,040
0,075
Planta
0,040
0,040
0,100
0,200
0,100
Corte
0,240

vimento misto em grelhas de enrelvamento de betão pré-fabricadas.

Laje de mármore branco 1,76x0,33x0,04m - Peça tipo A4
0,040
0,340
Estrutura dos degraus
(incluída no Projecto de Estruturas)
Geotêxtil (320g7m2)
0,330
0,150

Escadas intermédias do passadiço.

项目位置：葡萄牙科英布拉
客　　户：Coimbra Polis
占地面积：16 000m²
工程日期：2004 年～ 2006 年
预　　算：130 万美圆
景观设计：PROAP
细部设计：Gonçalo Byrne Arquitectos

Location: Coimbra, Portugal
Client: Coimbra Polis
Site Size: 16,000 m²
Project Dates: 2004~2006
Budget: US$1.3 mil.
Landscape Architect: PROAP
Detail Plan: Gonçalo Byrne Arquitectos

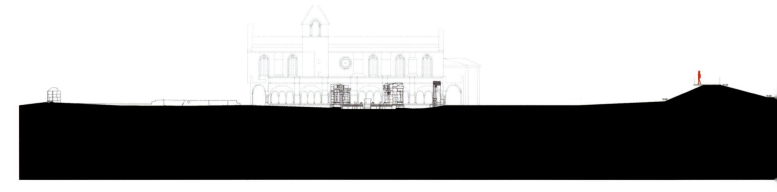

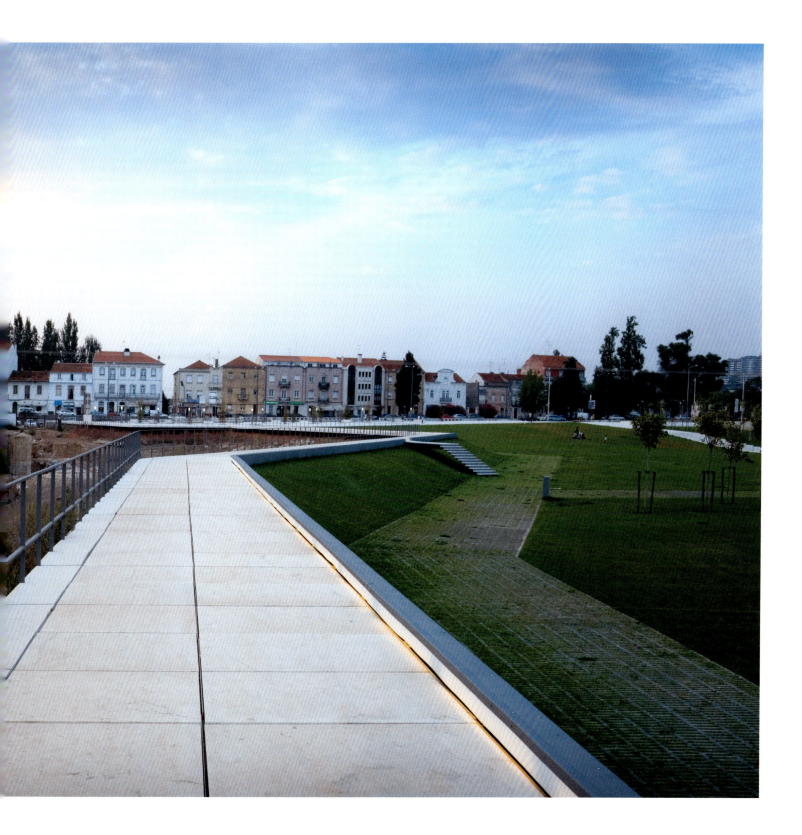

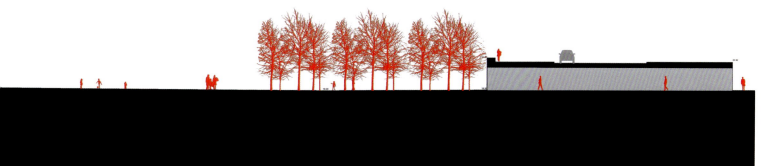

沉浸式设计——巴伦西亚动物园

Immersion Design—Bioparc Valencia

翻译　刘建明

巴伦西亚动物园分为两部分，其中一部分位于图里西亚河东岸，占地面积 100 000m²，另一部分位于图里亚河西岸，占地面积 50 000m²。这里曾是一片植物园，设计从零开始，改建后的景观将具有很大的发展潜力。

自然科学工作者、生态环境保护者以及动物学家、杜瑞尔野生动物保护协会和泽西动物园（Jersey）的创始人 Gerald Durrell（1951 年～1979 年）认为动物园最大的敌人是建筑师和兽医，该项目的设计团队（包括景观设计师、建筑师和兽医）则试图改变这种观念。设计师遵照了美国建筑师团队 Jones & Jones 所开发的"沉浸式动物园"的设计原则，该

原则已被运用到一些项目中，如西雅图森林公园（Woodland）的总体规划、位于亚利桑那州图森的索诺拉沙漠博物馆以及纽约布朗克斯动物园的丛林世界。

该项目的几个关键因素包括营造适宜动物生存的地形、在动物栖息地周围的植被布局艺术效果，以及在改造过程中将该项目与城市肌理有机结合起来。在历时 4 年的建造过程中，设计团队设计了 3 处独特的自然栖息地：马达加斯加岛、赤道非洲和非洲热带草原。

马达加斯加岛：这处栖息地所面临的难题与这个印度洋岛国因为砍伐森林而导致的生态系统危机有关。在几百万年

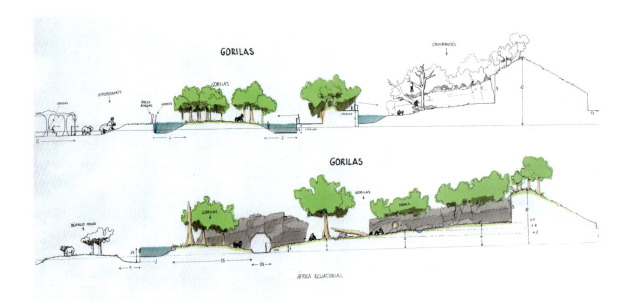

前，马达加斯加岛与非洲大陆分离，形成了独立的动物群和植物群。在解说中心外是一条被树木和悬于头顶的枯树所环绕的小路，马达加斯加岛特有的灵长类动物——狐猴，自由地穿梭其间，与游客直接接触，而砍伐森林对它们造成了很大威胁。

赤道非洲：这部分占公园面积的1/4，始于非洲村的小路一直伸向丛林，这里的植被非常茂密，以至于人们行走其间只能看到前方几米内的事物。游客先是到达原始热带雨林，然后来到生活着泽羚、河马和山魈的河岸，沿小路继续前进会看到一些生活在倾倒腐烂的枯树中的动物，如加蓬蝰蛇、蓝霓羚和小如兔子的羚羊物种。丛林越来越密，最后是赤色的水牛和豹子的栖息地，经过此处的人们可以看到猩猩，在另一端是河马和水禽栖息的红树林。

非洲热带草原：离开非洲村之后，热带草原区的小路立刻伸入被阿拉伯橡胶树和灌木所围绕的天地之中，这里生活着斑马和一些不同品种的羚羊。沿着溪边小路前进，游客

可以看到生长在水边的植被和典型的动物群，还可以看到一些鸵鸟在草原高处漫步，出口是开放式的白蚁土丘的造型，人工复制的地下动物巢穴反映出地下生物奇特的生活习性。小路变得越来越陡峭，不久便来到了地势较高的花岗岩地层———种称作"小丘（kopje）"的生态系统，这里是非洲热带草原特有生物猫鼬的栖息地。从这里可以俯瞰地势较低处的斑马和羚羊，这也是大型食肉动物被吸引到小丘之上的原因。人们还没意识到就已经进入了狮子栖息的中心地带（在确保安全的前提下），这些大型的猫科动物在岩石中休息。最后，经过长颈鹿和大象的栖息地，小路终止于茂密的棕榈树林中。

游客全程都沉浸在各种栖息地中，沿途经历着不同的地形，享受场地上的风景和高耸入云的岩层轮廓。海滩上的沙子被大型动物移走，形成了颜色深浅不一的土壤——土壤上不仅栽种有当地植物，还有许多可以作为园内动物的食物来源的非洲植物，如猴面包树等。

The City of Valencia, in Spain's Mediterranean coast, is still celebrating the opening of Bioparc Valencia, a new generation zoo managed by Rain Forest Valencia and designed by Rain Forest Design (RFD). This Madrid-based practice kicked off its specialized activity with the design and rehabilitation of Fuengirola Zoo, in Malaga, between 1998 and 2000, when the new park opened its doors as a new industry hallmark.

Built from scratch, Bioparc Valencia is located on 10 hectare of land on the eastern bank of the Turia River and 5 hectare on the western bank, where vegetable gardens grew not so long ago. The landscape possibilities were endless.

Twentieth century zoos – with noble exceptions – tended to be cold, denaturalized sites, often criticized for the methods used to exhibit living collections. The naturalist, conservationist and zoologist Gerald Durrell (1951~1979), founder of the Durrell Wildlife Conservation Trust and the Jersey Zoo, noted that the great enemies of zoological parks were architects and veterinarians. RFD put together a team of landscape architects, architects and vets to reverse the terms of this famous statement. Rain Forest works with the "immersion-zoo" principles developed in the 1980s by the American architectural practice Jones & Jones, and epitomized in key works such as the innovative Seattle Woodland Park master plan, added to by Mer Larson's later hyperrealist application of nature for the purpose of recreation (known as "thematization"), and best exemplified by the Sonora Desert Museum in Tucson, Arizona, or Bronx Zoo's Jungle World, in New York.

The topographic landscape demands of keeping animals in captivity, combined with the artistic effort of creating accurate vegetation layouts in their enclosures, while simultaneously integrating the project in an urban fabric in the process of rehabilitation were the key departure points for the project. Four years of construction enabled RFD to respond to three distinctly unique natural habitats: Madagascar, Equatorial Africa and the African Savanna.

Madagascar: This habitat problematizes issues connected to the endangered ecosystem in this island-nation in the Indian Ocean, plagued by massive deforestation. It's separation from Africa millions of years ago resulted in singular fauna and flora. Just out of the interpretation centre visitors can access a path surrounded

by trees and fallen trunks cantilevered above their heads, where different species of lemurs – the primates endemic to Madagascar, endangered by deforestation – roam freely in direct contact with the public.

Equatorial Africa: Occupying ¼ of the park, the path departing from the African village heads out to the jungle, and soon becomes a trail surrounded by exuberant vegetation so thick that only a few meters are visible ahead. The visitor reaches primary rain forest woodlands, and further down a riverbank populated by a band of sitatunga antelopes, kept company by a hippopotamus and a group of mandrills. Back to the trail, other species live inside a fallen, "rotten" tree trunk: the Gabon viper, the tiny blue duiker, and an antelope species as small as a rabbit. The jungle closes further, finally giving place to a clearing inhabited by red buffalo and a leopard. Visitors cross the clearing inside another fallen tree, from where they are able to observe chimps. The opposite end opens up to a mangrove habitat packed with hippos and water birds.

Ethiopian Savanna: The savanna route leaves the African village and is immediately surrounded by acacias and tall shrubs, amid which live zebra and different species of antelope. Following the path of a creek, the visitor finds riverside vegetation and its typical fauna. An underground animal dwelling replica reveals the curious life of underground species. The exit is done through an open "termite mound". Outside, among other animals, a few ostriches roam in the high grass. The path now becomes steeper, and soon granite formations rise in the grassland, an ecosystem known as kopje, which is home to unique savanna species such as mongoose. The path rises out of the savanna affording panoramic views over the zebra and antelope down below – precisely the reason why large predators are attracted to the kopje. Before the visitor realizes it, the path is in the heart of the lion habitat, where

the large felines rest in rock formations, with their prey safe on the background. The savanna path finally passes the giraffes and elephants, ending in a dense palm tree habitat.

Visitors are immersed in the habitats all along, exploring the topography inch by inch, and hence enjoying the scenography of the site and the dramatic silhouettes of rock formations rising towards the blue sky. Sand from the beaches is carried away by large animals resulting in polychromatic variations amid the vegetation – a combination of locally sourced species and African species fundamental to the animals' wellbeing, among which the flagship baobabs.

项目位置：西班牙巴伦西亚	Location: Valencia, Spain
客　　户：巴伦西亚 Rain Forest	Client: Rain Forest Valencia, S.A.
预　　算：6000 万欧圆	Budget: €60 million
占地面积：108 000m²	Surface Area: 10.8 ha.
项目时间：2003 年～ 2008 年	Project Dates: 2003 ~ 2008
景观规划：Rain Forest 设计公司	Landscape Planning: Rain Forest Design, S.L.
主要景观设计师：José Maldonado Castillo	Lead Landscape Architect: José Maldonado Castillo
首席建筑师：Luís María Ortiz Valero	Lead Architect: Luís María Ortiz Valero
兽医顾问：Gonzalo Fernández Hoyo	Consulting Veterinary: Gonzalo Fernández Hoyo

现实与幻想的融合

Intergration of Reality and Imaginary

翻译　董桂宏

胖困扰的儿童悉心交谈，充分了解这些儿童患者的需求。在该项目中，布林医院象征着现实世界，而兰博里特森林则象征着幻想世界，二者之间是此处与彼处的关系；而该项目旨在成为现实世界与幻想世界之间的衔接、此处与彼处的会合之地。该项目的空间开阔，为从不同角度欣赏景观提供了有利条件。

天然场地中的景观不免有"杂乱无章"之感，而该项目却展现出明显的几何图案景观。场地内的草坪呈阶梯状向森林方向延伸下去，在每一级草坪阶梯的边缘均建有坚固的挡土墙。从高处望去，仿佛是一级一级的矮墙试图挡住猛冲向下的草坪大军，而一旦草坪越过这一级一级的矮墙，便会投入到茂密的森林的怀抱。从高处向低处望去，可以看到每一座挡土墙均留有开口。该设计有两处精巧之处：其一，当游人从高处走向森林时会发现草坪阶梯之间虽有挡土墙阻隔，但是通过预留的开口，每一个草坪阶梯都相互连接，在空间上具有连贯性；其二，从森林的边缘向上望去，挡土墙与草坪阶梯的布局酷似金字塔的造型，威严壮观。

该项目的视觉效果非凡，从高处和从低处望去会有完全不同的视觉体验。另一个设计特色是修剪掉树枝的杂枝，这些树枝涂刷着白漆，看似散乱地伫立在场地中，但是在特定的角度可以看到它们构成了简单的几何图形。在顶端和底端的挡土墙旁设计师设置了一些这样的树枝，围合成半圆形。底端的一座挡土墙位于顶端的两座挡土墙的中央线上，在空间上构成三角形。设计师也有意使部分树枝形成三角形，与挡土墙的造型相呼应。这种视觉游戏很容易被人理解，游人也可以从不同角度欣赏这些独特的几何图形景观。

场地在白天与黑夜展现出完全不同的视觉效果。设计师在部分树枝上涂刷了夜光颜料，这种夜光颜料会在白天吸收太阳光，然后在夜晚发光，从而使树枝形成与白天相反的三角形造型。

设计之初，设计师对场地进行了细致的考察，充分利用了场地的现有条件。该设计的目标是深度挖掘原有场地的潜力，将原有的景观元素加以改造和重组，使场地从不同角度展现出迥然不同的景观效果。设计师将场地原有的景观元素以不同的方式重新配置，使其呈现出非凡的视觉效果。

该项目地处布林医院与兰博里特森林之间的过渡空间。法国 Art dans la Cité 协会委托 Jean-Pierre Brazs 公司设计该项目。布林医院是一家儿童康复医院，专门为 17 岁以下的儿童提供医疗服务。在进行设计之前，设计师与长期在布林医院接受治疗的儿童患者进行了交流，特别是与那些饱受肥

Work on site usually starts by seeking different points of view for the intervention, and then identifying the materials found on the site that can be used in the project. These interventions can be described as a manipulation of materials that have been extracted, displaced, transformed and replaced in a different order. The objective is to disturb a site, giving it an unexpected quality that still seems appropriate to it. The work is, therefore, born from and woven into the place itself.

The Remontees de Bullion project was the result of a request by the French association Art dans la Cité, who invites artists to intervene in a hospital setting and create ephemeral (or long lasting) works with the participation of hospitalized children. The pediatric and rehabilitation hospital of Bullion welcomes children up to 17 years of age and is located near Paris, in the heart of the Rambouillet forest. The children that participated in this project were being treated for obesity problems. The intervention site was located between the hospital (the everyday reality) and the forest (the imaginary). The setting provided ample opportunity to play with the relationships between real/imaginary, here/there, and with spatial movements and different viewpoints.

The space was a clearing, offering a strong geometric counterpoint to a natural and "disordered" world. Terraced lawns held in by solid retaining walls descended towards the forest. From above, it was possible to see a small wall seemingly halting the advancing lawn. Beyond it is the forest.

Moving forward, the opening in the wall too became visible. First surprise: the first expanse of lawn gives way to a second one and then to a third one as the user approaches the forest.

Second surprise: looking up from the forest edge, the user notices that the walls and stairs resemble a pyramid.

The first choice was to select different points of view that could be seen from an area above the lawn. Two views were chosen corresponding to specific alignment effects of the walls. Next, stripped branches painted white were planted in apparent disorder. This layout produced simple geometric shapes when viewed from the selected viewpoints. A semicircle of branches appeared where the bottom and top walls aligned. Here, a bottom wall was centered between two top walls.

A triangle of branches appeared where the walls form a triangle. Visitors understanding the visual game, search for the view from which they will be able to see the circle or triangle.

At night, opposite shapes were created by painting the branches with phosphorescent pigments that absorb light during the day and then glow at night, resulting in a circle within the triangle and vice versa.

项目位置：法国
预　　算：9200 美圆
占地面积：约 3600 平方米
项目时间：2007 年 5 月~6 月
景观设计：Jean-Pierre Brazs 公司

Location: France
Budget: US$9,200
Site Size: approx. 3, 600 sqm
Project Dates: May~June 2007
Landscape Design: Jean-Pierre Brazs

低成本景观——桑德大道

Low Cost Design—Sønder Boulevard

翻译　董桂宏

桑德大道纵贯哥本哈根市内城的韦斯特博露区
(Vesterbro)，建造之初旨在打造一个繁华的大都市面貌。虽然，
如今的桑德大道依然保留着 19 世纪的欧洲街头风格，但与其
当初的建造宗旨几乎相悖——这里全无繁华大都市的影子。每
天平均有 2700 辆机动车、1600 辆自行车和电动自行车途径桑
德大道，道路上喧嚣嘈杂，其主要的休闲区域则成了宠物的天
地，但却欠缺城市基础服务功能。考虑到当地居民的需要，设
计师开始大刀阔斧地对其进行改造。

1990 年以前，桑德大道上每 5.5km 就种植着上百棵树木，
这些树木将桑德大道划分为两个平行的区域。由于感染了荷兰
榆树病，这些树木到 1990 年已所剩无几，最后该区域只剩下
698 块巨大的花岗石，这使桑德大道成为了纯粹的交通通道，
没有任何风景可以让人流连。因此，在进行设计时，设计师反
复思考如何才能将桑德大道改造成一个受公众欢迎的公共空
间，以吸引当地居民到此休闲！

设计之初，设计师组织了很多次讨论会，广泛地邀请当
地居民参与讨论，以便深入了解他们对于未来的桑德大道有哪
些期望和要求。当地的居民也非常配合，并积极地表达了各自
的意愿。最终的讨论结果是将桑德大道改造成具有丰富活动
的休闲空间，让所有人都可以在此享受一份安宁与悠闲。因此，
设计师面对的挑战就是如何使桑德大道惠及所有人。基于这
一讨论结果，设计师将原有空间细划成众多分区，包括体育活

动区、游乐区、宠物专区、烧烤区、户外咖啡区以及小公园内
的冥想区，并在所有区域内都铺设了草坪。如有需要，各个分
区的功能还可以进行调整和变换，简单的场地设计使得这种变
换变得方便易行，这样的设计使桑德大道各种功能齐全又具有
灵活性。

设计师虽然只是小幅度地调整了场地原有的高差，但却
强化了场地的空间感。改造之前，桑德大道不过是连接 A 点
与 B 点交通的通道，而现在多样的植物和精巧的道路设计使
桑德大道摆脱了单一的通道形象。设计师将过去被中央区域分
隔的两个平行通道合二为一，并种植了更多树木以丰富场地空
间。场地的空间布局很简单，各分区内成排地种植着经过精心
挑选的树木，这些树木的发芽、开花和落叶的时间各不相同，
保证了一年四季景观的多样性，使人们在城市中心就可以欣赏
到美丽的自然景观。周边的建筑仿佛成为了它的框架，人们可
以从多个角度欣赏桑德大道的这种美感。

该项目改造可谓是众多城市空间改造项目中的典范。其成
功的标志就是在振兴城区的同时惠及当地居民，而不是将原有
空间特殊化、小众化。该项目不但具有经济效益和社会效益，
还为当地居民提供了更多的生活乐趣，共同将原有的场地改造
成属于每一位居民的新桑德大道。另外，该项目也是一项成
本较低的改造项目，其每平方米的造价在哥本哈根所有的改造
项目中是最低的。

Boulevards still have an air of 19th century Europe with its flaneurs, scandal and flirtation. But how does a contemporary boulevard actually function, where 2700 cars and 1600 bicycles and mopeds pass every day, and where recreational areas are mostly realms for dogs to relieve themselves in? Sønder Boulevard cuts through the inner city quarter Vesterbro like a fragment of the 19th century's dream of the great metropolis. As the years have passed there was, however, little left of the original visions. For this reason, the boulevard has been revitalized and adjusted to present needs.

Up until 1990 Sønder Boulevard consisted of 11/2 kilometer of hundreds of trees, which functioned as a barrier dividing the boulevard into two parallel parts. Following the arrival of the Dutch elm decease the only thing left was 698 granite boulders. There was really no reason for stopping by unless you had a dog to take out for an airing. It was an area of transit, with nothing else of any real interest on offer to the local citizens. So what was needed in order to turn Sønder Boulevard into a place worth spending time at?

To change this situation the local citizens were involved in numerous workshops where they expressed their many and often diverging wishes for a future boulevard adapted to contemporary needs. It was to be a space for everyone, full of different activities. SLA's challenge was to make the boulevard relevant for current and future citizens. Based on the workshop results we created space for sports and games, play, dog trots, barbeque, out-door cafes and meditation in small garden spaces. The boulevard is divided into a number of fields, which the citizens are allowed to change and use for whatever activity they may want to pursue. This simple configuration of the space allows for great variation. The urban space is robust yet also flexible with respect to changes, allowing future programs to be incorporated. Sønder Boulevard is a space in constant change. The levelling is slightly accentuated, only a little but enough to stimulate the sensation of space. A varied vegetation and complex spatial flow make it something different than the monotony of the sidewalk, which primarily accommodates transit from A to B. Cutting traffic down from two to one lane on each side of the central strip, gave us extra space for new functions which integrate with the new trees. The buildings work as a frame as the strip visually stretches from façade to façade allowing for views across the green areas of the boulevard. A simple arrangement of the space with rows of different tree species along the entire boulevard provides variation. The trees are chosen so they flower, spring into leaf and drop their foliage at different times. Thus the boulevard is always experienced in different ways, and draws the beauty of nature right into the inner city.

SLA's revitalization project at Sønder Boulevard is a demonstration projects in the municipality's strategy of developing new urban spaces. The criteria of success at Sønder Boulevard was to revitalize this local space without driving away the locals, rather preserving and reinforcing what is special and authentic in the area in collaboration with the local citizens. Sønder Boulevard is an upgrading of a local environment. It has benefited the local context in both social and financial terms as well as in terms of the amount of pleasurable experiences in the area. It is, furthermore, a low cost project with the lowest square meter price of all the recently modernized urban spaces in Copenhagen.

项目位置：丹麦哥本哈根
客　　户：哥本哈根市政府
预　　算：310 万美圆
占地面积：16 000m²
项目时间：2004 年～ 2007 年
建筑设计：SLA
照明设计：Hansen & Henneberg
　　　　　Lighting Engineers

Location: Copenhagen, Denmark
Client: Municipality of Copenhagen
Budget: US$3.1 million
Surface Area: 16,000 sqm
Project Dates: 2004-2007
Landscape Planning: SLA
Lighting Design: Hansen & Henneberg
　　　　　　　　Lighting Engineers

天然游乐场——渥尔比公园

Nature Playground—Valby Park

翻译 董桂宏

该项目始建于20世纪30年代,占地面积约为40万平方米,这里原是一座垃圾场,现已被改造成为哥本哈根市第二大公园。该项目是一座滨海公园,靠近凯尔夫波德(Kalvebod)沙滩,地理位置得天独厚。生活在都市里的现代人在紧张的工作之余,将充斥于大街小巷的广告、拥挤的交通、恼人的噪音、烟雾和粉尘等留在身后,来到沃尔比公园休闲放松,领略一望无际的自然风光。

该项目的改造目标是建设一座现代公园,其内部空间的层次清晰而连贯,并具有主要的标志性结构。项目改造的一个重要方面就是加强城市的生态功能,为自然景观留出广阔的空间。从1994年到2004年的十年间,对沃尔比公园的全面改造从未间断。1995年建设的青蛙公园标志着沃尔比公园迈上新台阶,此项目得到了WWF世界野生物基金会(WWF —World Wildlife Foundation)和图伯格基金会(Tuborg Foundation)的支持,旨在为青蛙与蟾蜍提供适宜的生存空间,并且为游客提供一种全新的生活体验。

该项目的草坪面积为70 000m²,草坪周围古树成荫。设计师新建了四座池塘(池塘的边缘建有石质堤岸)、一条蜿蜒的小路和一座4m高的小山,在小山上种植着两排樱桃树,白色的樱花缤纷浪漫、异常美丽,这些新增的景观也给草坪和古树注入了一股年轻的活力。为了增加游客的生活趣味体验,沃尔比公园允许游客在公园内种植野花,随处开放的野花像星星一样散落在公园中,与修剪平整的草坪形成对比。青蛙公园的周围新栽种了许多橡树,一条1000m长的大道横穿公园,大道两旁橡树成荫。池塘、点缀着野花的草坪和小山交织在一起形成了一道美丽的风景。1996年哥本哈根被评为"欧洲文化城

市",沃尔比公园为此相应地建设了17座圆形主题公园。

2001年,由景观设计师Helle Nebelong设计的沃尔比公园天然游乐场对公众开放。天然游乐场占地面积20 000m²,其施工均是由失业人员完成。

由于沃尔比公园过去是一座垃圾场,哥本哈根的环境部门要求施工时将原来地表0.5m深的旧土挖走,填培上干净的新土,但是环境部门又不允许将这些旧土弃置在沃尔比公园以外的地方,因此设计师利用这些旧土建造了一排小土丘,将天然游乐场与公园的其他部分分割开来。另外,游乐场外的天然

草坪也被视为游乐场的一部分。

天然游乐场由以下几部分有机组合而成：一排小土丘、铺满沙子和砾石的大型游乐区、绿色小岛屿、蜿蜒的小路、小村庄中呈弯曲造型的柳木小屋、折叠栅栏、野花区、蜗牛造型的大土丘（土丘螺旋形上升的道路将游人引向适合眺望的观景点）。设计师在游乐场内建造了一座 210m 长的圆形木桥，将整个游乐场串联起来。木桥高于地面 0.5m,桥面由橡树木板建成,所采用的橡树均是在荷兰榆树病风行时砍伐的树木。

景观设计师 Helle Nebelong 与丹麦设计学校的四名学生共同设计了该项目。他们在游乐场设计了 6 座小塔，其中的 5 座已经竣工。这些小塔是圆形木桥的重要节点，每一座小塔都有一个主题，分别是光之塔、风之塔、绿塔、鸟之塔和变幻塔。

在丹麦充斥着很多商业性的游乐公园，这种游乐公园在欧洲其他国家随处可见，而该项目的意义就在于为人们提供一处天然的游乐场。如今，该项目已经成为日间幼儿园、学校和学生课外活动小组的最爱，孩子们经常到此游玩。在周末，很多家庭都会来到这里休闲娱乐一整天。自 2001 年该项目对公众开放后，天然游乐场的很多设计细节也在不断更改，日臻完善。因此，游客每次来到这里都会有新的视觉体验，而且这个动态的过程从未停止。

Valby Park, with its 100 acres, is Copenhagen's second largest park and was established in the 1930s on a former dumping ground. The park lies by Kalvebod Strand beach and has the park qualities inherent in a location by the sea. It offers the stressed urban soul a chance to look out towards the infinite horizon, and relax without the constant disturbance of crossing pedestrians, obtrusive advertising, traffic, noise, smoke and dust.

The goal of the renovation was to create a modern park with a clear inner coherence and a primary identifying structure. An important aspect was to promote urban ecology and give nature greater latitude. From 1994~2004 the park has been totally renovated. The establishment of The Green Frog Park in 1995 marked the start of the new park. The project was subsidized by WWF World Wildlife Foundation and the Tuborg Foundation, for the purpose of creating suitable areas for frogs and toads, while offering visitors new experiences.

A 70,000 square metre grassy area surrounded by old trees has been given a special identity by the establishment of four ponds, a winding path and a four-meter high hill planted with double rows of white blooming cherry trees. Large stone banks slip along the edges of the oval ponds. Experiments were made with sowing wild flowers as a contrast to the trimmed lawn. A new wooden area of oaks was planted around the frog park, and a

1-kilometer-long avenue was planted across the park. A support area was laid out with water holes, meadows of wild flowers and hills. In 1996, when Copenhagen was European Cultural City, 17 circular theme gardens were constructed.

In 2001, a 20,000m² nature playground designed by landscape architect Helle Nebelong was finally opened to the public. The construction work was conducted as by unemployed people.

As Valbyparken is an old rubbish dumping area, the environmental authorities demanded that ½ meter earth had to be removed from the whole area and replaced by new, clean earth. Another condition set by the authorities was that the rubbish earth should not be removed from Valby park. Instead, it was built into a row of little mounds, separating the playground from the rest of the park. From the beginning of the design process the original woodland and the wide stretch of meadow outside the playground were identified as the spirit of the place.

The plan is made up of organically formed elements: a row of hillocks, large area with sand and gravel, small green islands, winding paths, a village of woven willow huts and plaited fences, an area with wild flowers and a very big snail-shaped mound with a path spiralling up it to a look-out point. The whole playground is pulled together by a circular 210m wooden bridge, which "floats" ½ meter above the ground. The planks in the bridge are from

the many elm trees, felled in Copenhagen due to a Dutch elm disease outbreak.

Landscape architect Helle Nebelong worked well together with four students from Denmark's design school. They designed six towers for the playground, of which five were constructed. The towers are placed as precise points on the circular bridge, each with its own theme: The Tower of Light; The Tower of Wind; The Green Tower; The Bird Tower; and The Tower of Change.

The ambition is that the playground should become a good alternative to the many commercial amusement parks, appearing everywhere in Denmark as in other European nations. The playground is now a favourite place for day-care centers, schools, and after-school clubs, who visit it on day trips. At weekends it is a very popular place for families to visit and stay for the whole day. Since the opening in 2001, Helle Nebelong has repeatedly intervened on the space, which thus became a continuous work in progress in which some detail is always changing. It is, thus, a dynamic work in progress without an end in sight.

项目位置：丹麦哥本哈根
客　　户：哥本哈根市
占地面积：20 000m²
项目时间：1996 年～ 2001 年
景观设计：Helle Nebelong，Sansehaver.dk
学生助手：Kirsten Due Kongsbach，Pernille Frank Dige，Pernille Bustrup，Fridrik Bjarnason
所获奖项：2006 年及 2007 年哥本哈根最佳游乐场（媒体评奖）

Location: Copenhagen, Denmark
Client: City of Copenhagen
Surface Area: 20,000 sqm
Project Dates: 1996~2001
Landscape Architect: Helle Nebelong, Sansehaver.dk
Design Students (Assistants): Kirsten Due Kongsbach, Pernille Frank Dige, Pernille Bustrup and Fridrik Bjarnason
Awards: Best playground in Copenhagen 2006 and 2007 (media award)

一墙一花园——盖布朗利博物馆

One Wall, One Garden—Quai Branly Museum

翻译 潘岳

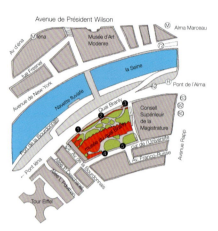

即使是在巴黎这样的大都市，一处地标性建筑的诞生也是非同寻常的。而将一栋建筑与有着非凡意义的植物园联系在一起就更为罕见。然而，要完全理解吉尔·克莱芒设计的植物园是如何将盖布朗利博物馆包裹住的，还需要对建筑内部进行仔细分析，并思考其中所蕴含的哲学问题和分歧。

盖布朗利博物馆是巴黎第一座大型博物馆。该博物馆并不是随意选址建造的，而是经过几个世纪的变迁才将河畔街区的居民全部迁走，在雄伟的埃菲尔铁塔边开辟出建造博物馆的空间。

博物馆的选址已经颇具争议，而为其筹集展品更成为一场公共关系的噩梦。巴黎两个主要的博物馆由于为盖布朗利博物馆提供了将近30万件展品而被迫关闭，这引起了民众强烈的抵触。盖布朗利博物馆渴望成为一所世界一流的研究机构，可以展现原生态艺术，反映非洲、亚洲、大洋洲和美洲等地的文明特色，或者呈现出与众不同的艺术形式。

设计师让·努维尔的建筑打破了规模、形态、颜色、尺寸和体量等方面的常规——立方体立柱从建筑的正面凸出来，从而为内部展厅开辟出特殊的空间；天棚高低起伏；入口坡道从一层的前厅开始，以环路的形式通向其他主要展厅。夹层提供了观察游人与展厅互动的机会；主要楼层的布置好似一处迷宫般的舞台。没有笔直的走廊，取而代之的是由貌似低矮的土砖墙形成的蜿蜒的小径，可能是设计师为了缅怀远去的非洲文明而刻意建造的。极简抽象派的灯光设计强调黑暗感和神秘氛围，铁制品总是零星而出其不意地出现在展厅中，看上去仿佛是陈列在玻璃橱中的图腾，使人眼前一亮。对于很多人来说，这一陈设暗含着原初性的理念。这是一座

不屑于解释的博物馆，来到这里的人们通常都会产生许多疑惑，但是不一定能得到所有的答案。在这里大人和孩子一样面对着未知事物，并且无法用语言去描述、解释。这也是一座没能摆脱殖民主义思想和"异域情调"束缚的博物馆。

毫无疑问，设计师让·努维尔的这一设计已经成为巴黎的一处地标性城市景观，设计师称其为一处"若即若离、似有却无"的景观。人们只单纯地认为这是该设计师的又一力作，而设计师自己却将其视为"所有景观元素都是为了唤起人们对原初性的共鸣的建筑""随意布局的梁柱都可能被错认为是树木或图腾""有实体的东西似乎消失了，使人们感觉到博物馆是一处隐藏于密林中、简陋的、没有立面的小木棚"。设计师说道："馆中所有的陈设都是为了通过柔和的差异来营造诗意——座巴黎式的植物园成为了一片神圣的树林，在其深处浮现出一座博物馆"。

这里的确变成了一片神圣的树林。巴黎设计师 Acanthe Paysage 和景观设计大师吉尔·克莱芒在许多方面获得了设计师让·努维尔所没有的优越条件。他们得到了大约20 000平方米的土地，使整个博物馆被包裹在不同风格和意境的密林中。由于博物馆处于这一地区的中心地带，所以景观设计以谨慎、谦逊的风格为主，但其中也不乏巧妙的构思。这处容纳了多个建筑的景观地带由蜿蜒的小径相连而成，哪怕是主立柱的悬桁下也是如此。两三条小路平行延展，有数米间隔，或横或纵，却常常被高的灌木和青草所掩盖。其中一条小路通向室外礼堂和附近的一处室内会场。每当夏季，这里便会有免费的公共演出。在小路的尽头有几条斜坡，由于节点和方向很多，容易使人们迷惑——那到底是连接两栋建筑

的人行道，还是表面为深褐色、用橘黄色元素分割空间的不同寻常的幕墙呢？

石板小路环绕在建筑之间，从低地青草和高山植物到小巧的英式庄园，从河畔植物带到草药种植区，从平坦的草坪到茂密的森林，共计169种新树、886种灌木和74 200种蕨类植物种植在这片用做集会和通道（许多参观者的确只是路过而已）的公共地带。在建筑下方的花园中，装饰着具有艺术色彩的设施——白天，半透明的丙烯酸棒隐藏在草坪中；夜晚，当埃菲尔铁塔的灯光作为背景闪烁时，它们则会呈现一种蓝紫色的荧光，这种景观装点方法极具趣味性和参考价值。

一条长220m、高12m的玻璃屏障将城市的喧嚣与河畔分隔开来。这道玻璃屏障也会带来另一种感觉——模糊了外侧城市景观和内侧草木的界限：汽车从玻璃屏障边疾驰而过，路人透过玻璃向内张望，树木抑或是树木的倒影使人们感到炫目而迷惑；园内的参观者也可以向街道望去，仿佛又回到园外。从上方的平台向下看，看到的景致与在建筑夹层中欣赏到的景致极为相似。

即使是在盖布朗利博物馆周边，景致也很有特色。一座政府行政大楼的外侧墙面对着河流和市区，据说这是世界上最大的植物绿墙。它是由世界闻名的垂直花园设计者Patrick Blanc与吉尔·克莱芒合作设计的，巨大的主立面上覆盖着150种、共计15 000株不同的植物，内部由另一面较小的植物墙作为补充。遗憾的是，由于缺乏特殊的系统以支持植物的根系、解决灌溉和排水等问题，这面植物绿墙已不再郁郁葱葱，但它仍然使路人惊叹不已，印象深刻。

It is rather uncommon for landmark buildings to see the light of day, even in a city like Paris. For such a building to come attached to a magnificent public garden is rarer still. Yet, understanding Gilles Clément's garden enveloping the Quai Branly requires a careful analysis of the building within, and a consideration of the philosophical intricacies and disputes involved.

The Quai Branly Museum (or MQB, as it has been nicknamed) was the first large museum to be built in Paris in a long time. But museums don't appear out of nowhere, and it took the flattening of a couple of centenary riverside blocks to make room for it, in a noble site neighbouring the Eiffel Tower.

If this was controversial, finding a collection became a public relations nightmare. The museum's aspirations were to become one of the leading world institutions featuring indigenous art from civilizations in Africa, Asia, Oceania and the Americas – or, generally speaking, "the Other". It took the closing (or severing) of two major museums in the city to supply Quai Branly with its portfolio of nearly 300,000 pieces. This was not without vehement opposition.

The Museum's stance is somewhat different. Jean Nouvel's building defies all rules of scale, shape, colour, size, and gravity. Cubic volumes spring out from the front façade, creating special exhibition units inside. Ceiling heights rise and drop at different points. An access ramp loops its way from the ground floor lobby to the main exhibition floor. Mezzanines give voyeuristic visitors the chance to observe others interact with the exhibition. The main floor is set like a stage, or a labyrinth of stages. There are

no straight corridors, rather sequences of curving paths lined by what seem to be low adobe walls, perhaps reminiscing of a long-lost African civilization. The minimalist lighting design reinforces the dark, mysterious atmosphere, in which loose icons appear in unexpected locations, displayed like totems within glass cases. For many, this setup implies a suggestion of primitiveness. For the museum, this is a place that defies explanations, where people come to be surprised but not necessarily to take away all the answers – a museum where grown-ups are as confused as children facing the unknown and lacking the language to decode it. This is, however, also a museum that has failed in its mission of avoiding the trappings of colonial thought and fascinated "exoticism".

It would clearly be an interesting exercise in understanding "otherness", if only the administration's stance on international debates was not so blunt. Following the position of the French state, the museum has exuberantly refused to return Maori warrior heads in its collection to the New Zealand tribe that claims their ownership for a proper burial. The episode only added to a long list of controversy surrounding this unique construction, but it also unmasked important issues of positionality.

For all the polemics, there is little doubt that Jean Nouvel's iconic design is already among the fundamental landmarks in Paris' cityscape, one with a "presence-absence or selective dematerialization", in the architect's own words. Where others see a simplistic understanding of indigenous architecture made

into a token showpiece in Nouvel's ever-growing portfolio, he sees a building "where everything is designed to evoke an emotional response to the primary object", in which "randomly-placed pillars could be mistaken for trees or totems" and one in which "what is solid seems to disappear, giving the impression that the Museum is a simple façade-less shelter in the middle of a wood". "All that remains", Nouvel says, "is to invent the poetry of the site by a gentle discrepancy: a Parisian garden becomes a sacred wood, with a museum dissolving in its depths".

And a sacred wood it became. In many ways, Paris-based Acanthe Paysage and master landscape designer Gilles Clément got right what Jean Nouvel perhaps did not. They were given almost 20,000 m² to wrap the museum in a lush forest of different styles and purposes. While Jean Nouvel's building is the centerpiece of the ground, the landscape discreetly upstages it through a variety of modest, yet intelligent artifices. The site includes multiple buildings are connected by meandering paths all over the landscaped area, even underneath the cantilevered area of the main volume. High shrubs and grasses often conceal the fact that two or three paths run parallel, separated by a few meters – horizontally or vertically. One of these leads to a great outdoors auditorium contiguous to its twin indoor venue. In the summer time, free public performances are put up, the inside/outside discrepancy disappearing when large windows are

wide open. At the end of the path, a variety of slopes confuse the eyes with multiple focal points and optional directions, be it the elevated promenade connecting two buildings, or the uncanny curtain wall painted in a dark ochre and punctuated by protuberating orange elements that stick out even from the distance.

Slated paths circulate from a niche to another, from the low moor grasses and alpine flora to the little English garden, from riverside vegetation to herbal compositions, from a plain lawn to a thick wood. Altogether, 169 new trees, 886 shrubs and 74,200 ferns and grasses were used to green-up this public space for intense use and passage (and many of the visitors are really only passing by). Underneath the building, where darkness takes over the garden, an art installation took over: during the day, what appear to be translucent acrylic sticks pepper the lawn; at night, when the top of the Eiffel Tower flickers in the background, they become a fluorescent blue-purple accent in the landscape, a playful landing strip that can be used as a reference.

A 220 m long, 12 m high glass palisade limits the urban sounds from the riverside thoroughfare. This glass barrier creates yet another layer of perception, blurring the boundaries between the outer cityscape and the inner greenery – cars dashing just outside the window, passersby looking in at the deceivingly dense woods, dazzled and confused by the sight or trees (or are

they blurred reflections of trees in the glass?), garden visitors looking out, taking no more than seconds to return to the outside if they wish. From the terrace above (all 2,500 m² of it), the bird's eye views of the garden mimic those that can be enjoyed from the mezzanines inside.

Even outside the perimeter Quai Branly makes a statement. The external, avenue-facing wall of one of the administration buildings faces the river and the city with what is claimed to be one of the largest vegetal walls in the world. Designed by the world master of vertical gardens, Patrick Blanc (in collaboration with Gilles Clément) the large façade includes 15,000 plants of 150 varieties, and is complemented by another – smaller – vegetal wall inside. Sadly, the living wall art installation has gone from healthy and lush to somewhat needy overtime, suggesting inadequate support systems for the plants' roots, irrigation and drainage. Visually, it is still as impressive for the astonished passerby.

项目位置：法国巴黎
客　　户：盖布朗利公共公司
占地面积：18 000 m²（花园）；800 m²（植物墙）
建成时间：2006 年
预　　算：3 亿美圆
建 筑 师：让·努维尔（2008 年普利兹克奖）
景观设计：Acanthe Paysage 和吉尔·克莱芒
植物墙设计：Patrick Blanc
照明设计：Yann Kersalé／AIK

Location: Paris, France
Client: Quai Branly Public Corporation
Site Size: 18,000m² (garden) 800m² (vegetal wall)
Completed Time: 2006
Budget: US$300 million
Architect: Jean Nouvel (Pritzker Prize 2008)
Landscape Design: Acanthe Paysage with Gilles Clément
Vegetal Wall Designer: Patrick Blanc
Lighting Design: Yann Kersalé/AIK

金色公园
Parco Dora

翻译：刘建明

都灵是一座具有悠久工业传统的城市，现在正面临着一场结构性的复兴。在这场复兴运动中有一个关键的项目"金色公园"(Parco Dora)（有时也称之为金色荆棘公园 Parco Dora Spina)，其选址临近传统的市中心地域。市政当局拟将被工业化毁坏殆尽的约 37 万平方米的地方改造成一个大型城市公园，使其成为都灵后工业化景观的代表之作。2004 年，Latz I Partner 联手复兴景观设计大师 Peter Latz 成功摘得了这一设计竞标，从而将欧洲范围内其他极具创意的设计方案远远地抛在了身后。

该项目选址为意大利原最大的汽车制造公司菲亚特及其供应商的厂址，眼下遍地是工业化后的疮痍，不时仍有醒目的工业化建筑映入眼帘。朵拉 (Dora) 河横贯该项目场地，几条街道将这一区域分割成五个特色各异的街区。

该项目的几大主要特色如下：

一、保存和变形——当前的工业化遗迹正是保存历史的见证。工业化遗迹将作为未来公园的一个部分，转型后即可通过新的方式来注入新的价值，创建新的内涵，进而丰富未来公园的主题定位。

二、特色景观元素的交叉互连以及与临近居民区的衔接——一座步行桥跨越朵拉河两岸，连接南部主入口与冷却塔附近的草甸公园，中央展示区内部隐藏着原来的生产车间和钢筋梁柱"丛林"。空中走廊是另一大关键的连接方式，它引导着游客从西部草甸、水景花园穿越中央地带通向东部地势较高的区域。空中走廊的平均高度约为 5 米，提供了另一种位面感知体验，并可使人们安全地跨越各个街区。适于步行的小树林、步行道与花园网络，增强了主通道之间的连接，并突出了公园主题。

三、朵拉河完美地融入公园概念——在"都灵——水的城市"这一城市开发项目的过程中，朵拉河得到了彻头彻尾的改造。河岸的翻修是该项目的基本任务之一，完工后游客和市民方可再次通过河岸亲近朵拉河。之前掩盖朵拉河的水泥板都被清除掉。朵拉河得以再次焕发新颜，成为公园的动态"脊梁"，人们可以重新在河道沿线尽情地放松娱乐。

四、雨水与暴雨径流管理——生态水流系统应当包括雨水的技术性收集和保存，完美地利用并能为游乐设施提供水源。屋顶和密闭表面的径流汇入开放的水渠，为花园和游乐设施提供水源。盈余的雨水将保存在水塔中或注入朵拉河。

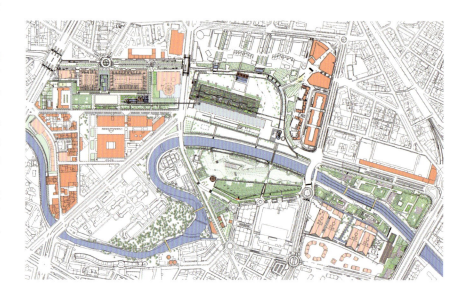

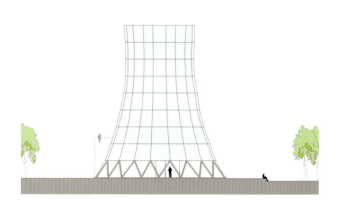

Torino, a town characterized by a long industrial tradition, passes through an ongoing structural renewal. One of its key projects will be the "Parco Dora" (sometimes referred to as Parco Dora Spina): In close vicinity to the historic city centre. The 37 hectares of formerly devastated land will be transformed into a large city park, transforming a postindustrial landscape are in Torino. Latz + Partner, with Peter Latz – one of the pioneers in this kind of intervention – at the helm won a very competitive bid for this significant project in 2004, leaving behind many other interesting proposals by some of the most creative practices across Europe.

Once the location of a car factory for FIAT, Italy's largest automotive corporation, and its supplying industries, the site is now widely demolished but still possessing some impressive industrial structures. The river Dora runs through the site, which is also intersected by several streets, dividing the area into five distinct sectors.

The main strategies of the project were as follows: (i) preservation and metamorphosis – existing traces of the industrial history shall not deny their origin. Preserved as an industrial heritage, they will be layers of the future park. By transforming their character,

they gain new value and will be used in new ways, creating new connotations, which will in turn warrant future identification with the place; (ii) interlinkage of characteristic elements and connection to the adjacent town quarters – a pedestrian bridge crossing the river Dora will connect the southern main entrance and the meadow park around the cooling tower with the central event area inside the covered former production hall and the "grove" of the steel pillars. An elevated walkway represents the second more important linkage. It leads the visitor from the western park side with the water gardens through the central area to the higher lying terrain in the east. Running on an average height of approximately five metres above ground, it offers a second level of perception and secure traversing of the street. Groves with completely walkable surfaces, a network of promenades and gardens built into still existing sub-constructions and inside the walls of former buildings, will complement the main connections and park elements; (iii) integration of the river Dora into the park concept – in the course of the urban development project "Torino, Città d' Acqua" (Turin City of Water) the river shall be rediscovered for the town. Redeveloping the riverbanks, and making them available again for visitors and residents, is an essential task. The concrete plate covering the Dora will be removed. The river will thus gain new significance as the lively "spine" of the park, enabling public leisure activities on and along the reopened watercourse; (iv) rain and storm water management – an ecological water system should combine the technical collection and retention of rain water with its aesthetic use and pleasure in the park. Run-off from roofs and sealed surfaces is meant to flow in open channels, to feed water gardens and play facilities. Surplus water should be stored in cisterns or fed into the river Dora.

项目位置：意大利都灵
客　户：都灵市
占地面积：37 万平方米
预　算：3800 万美圆
项目时间：2004 ~ 2011（第一阶段于 2008 年完工）
景观设计：Latz ＋ Partner
其他参与方：STS S.p.A.（博洛尼亚）
　　　　　　Archictect C.Pession（都灵）
　　　　　　Artist U. Marano, Cetara
　　　　　　G.Pfarré Lighting Design（慕尼黑）

Location: Turin, Italy

Client: City of Torino

Site Size: 37 ha.

Budget: US$38 million

Project Dates: 2004–2011 (1st phase finished 2008)

Landscape Architect: Latz ＋ Partner

Other Credits: STS S.p.A. (Bologna)

　　　　　　Archictect C.Pession (Turin)

　　　　　　Artist U. Marano, Cetara

　　　　　　G.Pfarré Lighting Design (Munich)

形式与功能的完美结合——Adnams旗舰店与咖啡厅

Perfect Combination of Form and Function—Adnams Flagship Store and Café

翻译　刘建明

Adnams
维多利亚街，索思沃尔德
市场街
2006 年 2 月 6 日

　　该项目建在索思沃尔德市中心中部的废弃工业用地上，包括一个公共广场，每周在此举行一次农产品交易。

　　2002 年，Ash Sakula 公司在该项目的设计竞标中胜出。当时，Adnams 刚刚将配送中心搬至索思沃尔德城外，该项目棕地得以重新进行开发，而这只是棕地复兴宏伟规划的第一期。后期开发项目包括 34 个新的住宅项目和连接至 Tibby's Green 以及邻近文物保护建筑——圣埃德蒙教堂的步行道。该项目规划范围内将建设 10 个低成本住宅社区。

　　公共空间是该项目的核心特色之一。长久以来，由于被用于工业，该项目棕地几乎与公众隔绝。原配送中心库房的规划不合理，影响了公共空间的质量。新的集市广场位于该项目的核心部位。沿着新的葡萄酒店从维多利亚大街就可以进入集市。北部的住宅项目完工后，新建的步行道将连接邻近的教堂和 Tibby's Green，而此前二者是被隔断的，因此未被充分利用。由此，索思沃尔德闭合的街巷网络和公共空间与赫赫有名的繁华商业街 High Street 完美衔接。住宅区后面私密空间的布局紧凑，符合高密度住宅区的特点，其中就包括了索思沃尔德亟须的 10 个低价住宅社区。当然，这与同公共绿地的完美衔接以及靠近新建集市广场是分不开的。

　　集市广场的地面是混凝土和萨福克 (Suffolk) 砾石的混合料。广场表面嵌有从最近的海滩防波堤维护工程中淘汰下来的木桩。硬质景观的外围种植着本地灌木和草种，突出一种随意、模糊的边界意义。这是索思沃尔德典型的半开放公共空间形式，也是索思沃尔德市民在寻求完全开放的公共空间过程中所引以为豪的成果。广场配置了自动跳起的电源插座，用于每周一次的广场农产品交易。照明设备形式随意，却引人注目。固定的户外设施是为了鼓励公众真正地利用广场作为交易场所。此外，集市广场上还设置有耐久长用的公共坐椅。

　　广场中央的水泥台周围是 3 棵成熟的悬铃木。广场上并没有种植很多树木，从而留出了充裕的游乐与坐椅空间。咖啡厅附近有一个固定式长椅。咖啡厅的建筑不高，如此设计是为了保证欣赏东面圣埃德蒙教堂的良好视线。咖啡厅的前后外立面都镶以玻璃，可以看到咖啡厅后面狭长的中式花园。

　　Ash Sakula 通过布置广场有限的空间，设计具有多种高度入口的单层店面结构以及满足其他通达性要求，完美地表达出包容式的设计风格。

　　商店本身是一个分为两部分的单层建筑。前半部有一个谷仓式的斜面屋顶，开放的山墙朝向维多利亚大街和新广场。山墙斜面的走向与 High Street 的方向一致。

　　商店后半部位置较低，从广场可以看到圣埃德蒙教堂。后半部包括许多零售区和一个咖啡厅，在温暖的季节甚至还可能延展至广场。咖啡厅设计了绿色屋顶，在夏季可以保持建筑内部的清凉。咖啡厅摆设了两个老式 Adnams 酿酒桶用做装饰，亦可作为小而整洁的静思处。

　　商店的主要建材为木材，再搭配少量钢框架。墙壁和斜面屋顶使用羊毛织物来隔热。玻璃窗装在橡木框架内。所有的玻璃窗都是高性能的双层玻璃幕墙——这在零售店的设计中并不常见。

　　整个建筑通过窗户和中央的采光屋顶来实现通风，如此一来可以避免采用机械制冷。同时，采光屋顶和玻璃外立面还能增加建筑内部的日光亮度，从而降低所需的人工照明成本。

The new Adnams flagship store and café is located on a brownfield site in the middle of the historic town centre of Southwold. The project includes a new public square for a weekly farmers' market.

Ash Sakula won the commission following a design competition in 2002, when site became available for redevelopment after Adnams moved their distribution centre to its new location just outside Southwold. A larger masterplan was drawn up, of which this project is the first phase of delivery. The next phase will include 34 new homes and new pedestrian links to Tibby's Green and the adjacent St. Edmund's church, a protected building. Ten affordable homes will be pepper-potted throughout the scheme.

Public space has been one of the core drivers of this project. For decades, the site had been closed to the public due to its industrial use. The old distribution shed was out of scale and offered no positive contribution to open space at all. The new market square is now located at the heart of the scheme. It is accessed from Victoria Street along the new wine store. Upon completion of the housing at the north, new pedestrian links will be opened up towards the adjacent churchyard and Tibby's Green, which had been cut off previously and under-used as a consequence. This way, the close-knit patchwork of Southwold's alleys and open spaces is continued all the way towards the edge of the historic High Street. The private spaces behind the houses are kept compact in line with the high density of homes, including ten affordable units, which are desperately needed in Southwold. This was made possible by the good connection to the public green, as well as its proximity to the new square.

The square surface is concrete with exposed Suffolk pebbles

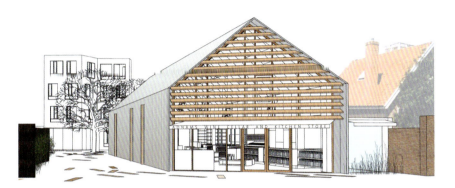

aggregate, carried out by a local contractor. In the surface are embedded, re-used timber piles salvaged from the recent beach groynes refurbishment in Southwold. The perimeter of the hard landscaping is planted with a mix of local shrubs and grasses, which provide an informal, soft edge. This is typical of semipublic spaces in Southwold, where people take great pride in looking after public space. The square is equipped with pop-up electrical sockets which will enable a weekly farmer's market to take place. Lighting is informal and inviting.

At the centre of the square are three mature London plane trees around a permanent concrete table. A few of the timber sections are raised here, providing informal playing and seating areas. A further fixed bench is located closer to the café. The café building has been kept low in order to provide good views towards St. Edmund's church at the east. The café façade is glazed front and back, giving a further view into the slim Chinese garden behind the café.

Ash Sakula made an inclusive design statement by creating a space limited to a square and a single-storey structure with level approaches to all entrances, as well as other universal accessibility features.

The store itself is a single-storey building in two main volumes. The front part has a barn-like pitched roof and opens up gables towards Victoria Street and the new square. The gables are skewed responding to the orientation of the High Street.

The rear part of the building is kept low to allow views from the square towards St. Edmund's church. It includes more retail space and a café, which will spill out across the square in the warm season. The café has a green roof, which helps keeping the building cool in summer. Two old Adnams brewery vats have been incorporated into the cafe facade and reused as snug retreats. The construction of the store is mostly timber, with a

Adnams

维多利亚街，索思沃尔德

市场街

2006 年 2 月 6 日

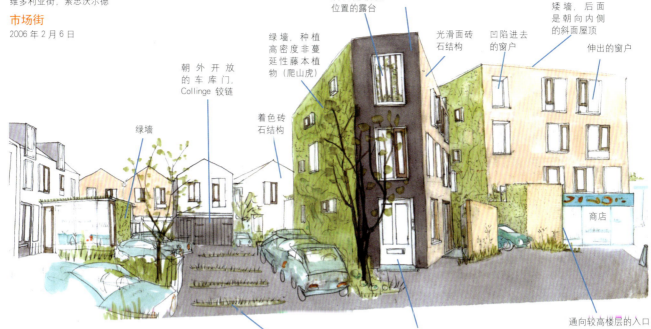

绿墙

朝 外 开 放
的 车 库 门，
Collinge 铰链

着色砖
石结构

绿墙，种植
高密度非蔓
延性藤本植
物（爬山虎）

开 放 窗 景，
后面是较低
位置的露台

喷有沥青的
砖石结构

光滑面砖
石结构

凹陷进去
的窗户

矮 墙，后 面
是 朝 向 内 侧
的斜面屋顶

伸出的窗户

商店

通向较高楼层的入口

经过 "Grasscretetype"
处理的海藻接缝

通向较高楼层的入口

minimal steel frame. Walls and the pitched roof are insulated using sheep's wool. Full-height glazing is framed in structural oak. All glazing is high-performance double glazed curtain walling – unusual for a retail unit.

The entire building is naturally ventilated through opening windows in the façade and a large central rooflight. This way, mechanical comfort cooling could be avoided. The rooflight and the glass façades also increase daylight levels so that lower artificial light levels can be used.

Reclaimed timber sections from the Southwold beach groynes are embedded in the market square surface between irregular concrete surfaces with exposed Suffolk pebble aggregate. The edges of the square are planted with local species grasses and shrubs, giving the entire development a soft edge, which is typical of Southwold's streetscape. Fixed outdoor furniture has been installed to encourage true public use of the square. Permanent public external seating is provided on the market square.

项目位置：英国伦敦
项目时间：2008 年 1 月～ 11 月
占地面积：5300m²
客　　户：Adnams PLC and Hopkins Homes
预算（店面与景观）：190 万美圆
景观设计：Ash Sakula 建筑师事务所
所获奖项：家居设计—2008 室内设计奖
　　　　　2009RIBA 奖项（提名）

Location: London, England
Project Dates: January 2008 to November 2008
Site Size: 5,300m²
Client: Adnams PLC and Hopkins Homes
Budget (store and landscape): US$1.9 million
Landscape Design: Ash Sakula Architects
Awards: Design for Homes - Housing Design Awards 2008
RIBA Awards 2009 (nomination)

别样的艺术——De Nieuwe Ooster墓园

Unique Art—De Nieuwe Ooster Cemetery

翻译　董桂宏

迄今为止，该项目是荷兰最大的墓园和火葬场，其占地面积为 330 000m²，包括 28 000 座坟墓。该墓园至今已有 117 年的历史，分三期建设而成，一期 1889 年建成、二期 1915 年建成、三期 1928 年建成。直到 2003 年，De Nieuwe Ooster 墓园已成为荷兰的标志性墓园。

墓园如同一面镜子，无论何时都可以清晰地反映出社会关系、葬礼习俗、集体与个体的关系，尤其能反映出人们如何看待自然与景观设计、建筑设计的关系。因此，该项目的改造不仅是一项景观实践，还是通过层次清晰的空间结构，表达出现代人对于死亡、葬礼以及痛失亲人的理解。基于以上考虑，设计师致力于创造出艺术价值与景观价值并存的墓园改造工程。

设计师认为每一个墓区都应该有其鲜明的特征，因此并没有将已有的三个墓区在空间上连接起来，而是与其恰恰相反。设计尽量求同存异，将其划分为三个界限分明的区域，以此强调每一个墓区的特性。

墓园的二期设计缺乏明显的特征，因此设计师大胆地创造出其独有的属性。在墓区的空间布局上不仅保留了原有公墓，还为未来需要新增的公墓预留了空间。随着社会等级观念的逐渐削弱，个体的想法和意愿得到了充分的尊重，因此设计

充分考虑了个体的意愿——希望百岁之后长眠于斯宾格曲径 (Springer's curving path)。

既要求设计多变，又要尽量满足这个共性需求，从而设计出非正式的空间结构，使死者能够安葬在斯宾格曲径 (Springer's curving path)。

第二次扩建后的墓区形成了相互平行的线形空间，设计师还在部分场地设计了一些树篱，有意地将空间分割开来。墓区内的骨灰安置所和池塘也各具特色。原有地葬礼区和缅怀死者的花园也是墓区的一部分，绿色的树篱将二者分割成独立的空间。另外，墓园内还稀疏地种植了一些白桦树。

65 处墓穴也是扩建后墓区的一部分。这里主要为公墓墓穴（使用期为 10 年），每一处墓穴可以安葬 5 口上下叠加的棺材。

65 处骨灰安置所位于相互平行的线形空间中，大部分原有空间的分界线予以保留。骨灰盒前后摆放，形成一条长长的直线，而骨灰盒封顶的木板略微高出地表。虽然这一设计的空间结构完全相同，但人们可以依据自己的意愿在骨灰盒封顶的木板上进行多种多样的个性化设计：木板本身设有凹槽，用于放置纪念性牌匾。

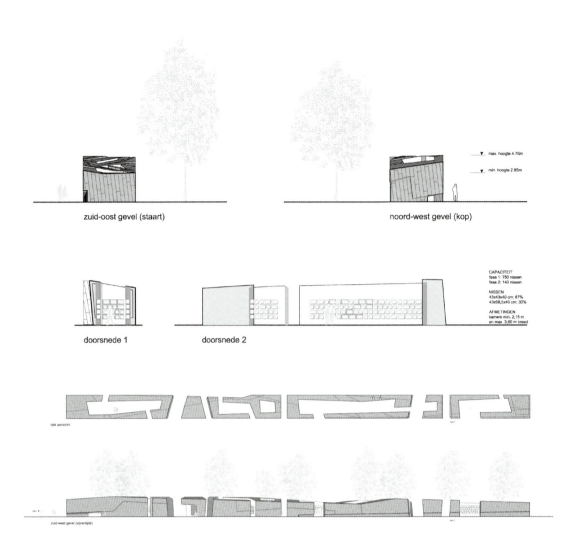

zuid-oost gevel (staart)

noord-west gevel (kop)

max. hoogte 4.75m

min. hoogte 2.95m

doorsnede 1

doorsnede 2

CAPACITEIT
fase 1: 750 nissen
fase 2: 140 nissen

NISSEN
43x43x40 cm; 67%
43x68,5x40 cm; 33%

AFMETINGEN
kamers min. 2,15 m
en max. 3,60 m breed

dak aanzicht

zuid-west gevel (vijverzijde)

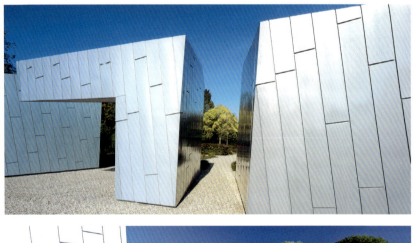

The site: De Nieuwe Ooster cemetery and crematorium in Amsterdam is by far the largest in the Netherlands. Measuring 33 ha. and containing over 28,000 graves, it was built in three phases, in 1889, 1915 and 1928. In the 117 years of its existence, it has undergone many changes. In 2001 a process of renovation and transformation process was started. Since 2003 it has been a national monument.

Cemeteries have always been mirrors of society, representing social relations, burial practices and the relationship between the collective and the individual, not to mention prevailing ideas on nature and developments in design and landscape architecture. For all these reasons, the design for the new cineraria in De

Nieuwe Ooster cemetery was not an isolated exercise: it required a clear spatial intervention, linked to a vision of how people currently approach bereavement, burial and reminiscence. The cemetery is also a popular park for other visitors, an aspect of the design that lies in close relationship with current views on death. The artistic and landscape value of the site undoubtedly contribute to this attraction. Rather than connecting the three different zones spatially, the designers believed each zone should be given its own identity. Enhancing the contrasts created a clear partition into three areas, thereby emphasizing the qualities of each individual part.

For the second extension, which lacks a single unifying quality, a

new identity has to be created. This will involve an intervention that is bold but relatively easy to implement. The new spatial structure of this zone shall accommodate the existing burial areas as well as create a framework for the necessary extension. Society becomes less hierarchical and more individualistic: everyone wants to be buried along Springer's curving path. People also have more divergent opinions, ideas and wishes.

De Nieuwe Ooster cemetery intends to meet such demands. There is thus a clear demand for diversity and to meet individual wishes within an informal spatial structure in which everyone can lie along the path.

The vision for this zone represents new developments within a new linear structure. Overall, the zone is underlain by parallel strips of various widths and design principles, some of which contain hedges that divide the zone into spatial compartments. The columbarium and the pond are special zones in this area. The existing burial areas and the garden of remembrance are incorporated within the zone as compartments with green edges. Silver birches are spread loosely throughout the zone.

Section 65 is also part of the second extension. For this section, a design was made for general burial chambers in which five coffins can be placed one above the other. Burial rights expire after ten years.

The design for section 65 fits within the vision for the strips

created by the parallel lines. Most of the existing boundary will be retained. The burial chambers will be constructed in long lines, one behind the other, their cover slabs raised slightly above ground level. Within a uniform spatial structure, the covers can be used for the expression of diversity and individual wishes: they contain recesses in which memorial tablets can be laid.

项目位置：荷兰阿姆斯特丹 Watergraafsmeer

客　　户：De Nieuwe Ooster begraafplaats，crematorium en gedenkpark

总 成 本：160 万欧圆

占地面积：20 000m²

项目日期：2004 年～2006 年（2011 年开始新一期工程）

景观设计：Karres en Brands 景观设计事务所

设计团队：Sylvia Karres，Bart Brands，Lieneke van Campen，Joost de Natris，James Melsom，Alejandro Noe，Marc Springer，Jim Navarro，Julien Merle，Pierre—Alexandre Marchevet

池塘设计：Maria van Kesteren

所获奖项：Topos 奖
　　　　　Torsoloranzo 奖的亚军
　　　　　Rheinzink 奖的冠军

Location: Watergraafsmeer, Amsterdam, The Netherlands

Client: De Nieuwe Ooster begraafplaats, crematorium en gedenkpark

Budget: €1.6 million

Site Size: 20,000m²

Project Dates: 2004~2006 (new phase in 2011)

Design: Karres en Brands landschapsarchitecten bv

Design Team: Sylvia Karres, Bart Brands, Lieneke van Campen, Joost de Natris, James Melsom, Alejandro Noe, Marc Springer, Jim Navarro, Julien Merle, Pierre-Alexandre Marchevet

Design Charons in Pond: Maria van Kesteren

Awards: Topos-Award
　　　　2nd prize Torsoloranzo Award
　　　　1st prize Rheinzink Award

城市标志景观——奥登斯科格交通环岛

Urban Identity—Traffic Junction at Odenskog

翻译 刘建明

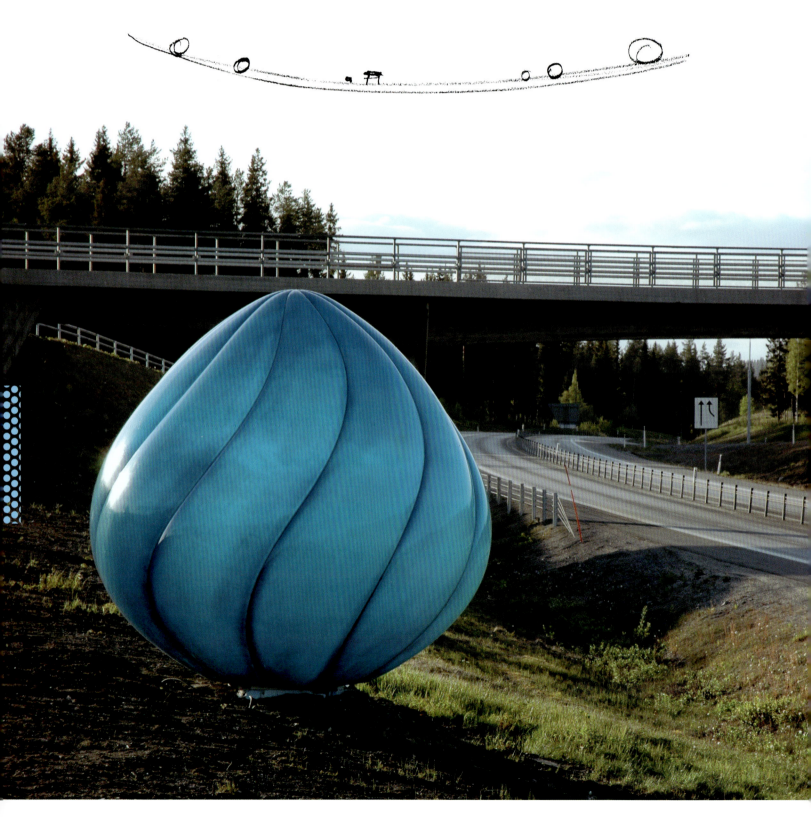

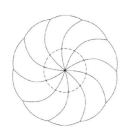

奥登斯科格交通环岛是从 E14 路进入瑞典厄斯特松德市的门户，这里也是通往瑞典北部地区的主要通道。厄斯特松德市是冬季一个重要的户外休闲地，因此从高速公路进入市区的地标性通道对于驾车前来的游客来说是至关重要的。这一区域的景观以密植的松树林和由小型农作物点缀的云杉林为特色。为了突出地域特点，环岛内的景观（直径 140m 的圆形范围内）以低矮的斜坡为主，并成排种植了唐棣。

6 座与过往车辆尺寸差不多的冷色系宝石蓝发光雕塑以不规则形式林立在环岛内，此布局更加突出了斜坡和唐棣林的不对称风格。白天，以玻璃纤维聚酯塑料为原料制作的雕塑完全不透明，给人以坚实稳固的感觉，尤其是和瑞典白色背景的雪地景观交相辉映，形成了更为独特的景观。到了夜间，这些雕塑摇身一变成为半透明的内部发光体。雕塑的色彩在不同季节为周围环境添姿添彩，无论是冬日的雪地景观还是夏日的郁郁葱葱。交通环岛的雕塑及精心设计的景观引起了国际景观行业的广泛关注和讨论。在很短的时间内，这一城市交通通道便为大家所认可，成为厄斯特松德市的标志性景观。

The traffic junction at Odenskog marks the entrance to the city of Östersund, Sweden, from Road E14. It is therefore a major gateway to the northern region of Sweden. The city of Östersund is a significant winter outdoor recreation destination, whereby the landmark of the entrance to the city from the highway is of great importance to all of the visitors entering by car. The landscape in this area is characterized by dense coniferous forest and mainly spruce interspersed with small-scale agriculture. To underline the significance of the place, the landscape within the circular junction – with a diameter of 140 m – has been shaped with low banks and planted with lines of Amelanchier spicata.

The asymmetrical pattern of the banks and lines is completed with six asymmetrically placed icy blue-towards-turquoise light sculptures roughly the same scale as the passing cars. In the daytime the sculptures – made of fiberglass-reinforced polyester plastic – are opaque and make a solid impression, particularly against the white background of Sweden's snowy landscape. At night they are semitransparent and lit from within. The color of the sculptures complements the surroundings over the different seasons whether it's the snowy landscape of the winter or the lush forest green of the northern summer. The traffic junction with its sculptures and shaped landscape has attracted a lot of attention and discussion around the country and beyond. Within a short time this urban gateway intervention alone has established Odenskog as a place with its own identity, which the city borrows as its own.

项目位置：瑞典厄斯特松德市 E14 路

预　　算：约 60 万美圆

客　　户：瑞典路政管理处

项目时间：2005 年~ 2007 年

占地面积：30 000m²（中心环岛 15 000m²，新树林 15 000m²）

景观设计：GORA 艺术与景观公司

Location: Road E14, Östersund, Sweden

Budget: approx. US$600,000

Client: The Swedish Road Administration Mitt

Project Dates: 2005~2007

Surface Area: 30,000 sqm (15,000 sqm roundabout, 15,000 sqm new forest)

Landscape Architect: GORA Art & Landscape

旧貌换新颜——汉堡港口新城

Urban Regeneration——Hamburg Hafencity

翻译　谷晓瑞　李沐菲

与市中心毗邻的水库区的南部曾是一个港口，对此地进行的一项重要改造就是港口新城西部公共空间的建设。随着时代的变迁这一地区也在不断发展，以满足港口和工业发展的需要；潮汐的交替塑造了港口所特有的盆地地貌。

在改造过程中，将中心城区新建的一些功能性建筑的地基提升了近3米，如居民楼、办公楼、商用大厦、文化中心和娱乐城等，以防止风暴潮侵袭。

为了将这个曾经供远洋货船停靠的港口改造成适宜人类居住的市区，设计公司设想将这里打造成一个充满生机且多层次的公共空间，这个公共空间将成为港口新城西部的枢纽区。

规划区域包括Sandtorhafen、Grasbrookhafen、港口的两端及其周边地区、周边新规划开发的广场以及港域沿岸的码头周边地区，并依照每个区域的自身特点进行设计。设计的亮点在于将处于不同水平线上的公共空间与私人空间、陆地与海水

巧妙地整合在一起。位于 Grasbrook 一侧的港口与整体设计方案相呼应,增添了很多时尚的景观建筑元素。曾被炸弹破坏的地方种上了垂柳,而台阶一直延伸至海边。这边的港口设置了很多观景点:两个高低错落的广场、一个户外餐厅和水边一座林木茂盛的小山。该区域景观设计的主题始终如一,一直延伸到码头的浮桥。

Sandtorhafen 将会成为货船停靠的港口。设计师设计了一座漂浮广场,浮桥的设计仍然延续以往的风格,与 Sandtorhafen 所在的台阶相连通。Sandtor 的港口修建了一片宽敞的台阶,一直延伸到海边并与 Sandtor 公园形成过渡。广场的地理位置优越,视野开阔,吸引了很多游客驻足欣赏。

设计降低了洪灾风险,使人们可以近距离地观赏潮汐,形成适宜人居的滨水环境。规划后,从居民区到海边都充满了活力,利用水和植物让人们尽情享受大自然艺术景观,人们可以与自然亲密接触。

巨大的浮桥平台是休闲的好去处,人们可以在这里乘坐小船或赛艇,独特的浮桥也给水面增添了一丝生机。从桥边和浮桥的孔隙中能看到水流,宛若一个池塘。人们可以在步道上悠闲地眺望海边。这里每年平均会有 2 次~ 3 次大暴雨的天气,只有此时,这里才会受到洪水的威胁。在街道层设计了漫步小路和休闲区,将拥堵的机动车道和步行道分开。

该项目的最大特点在于将不同水平面上的坡道、台阶、吊桥连接在一起,并构成一个整体。各种各样的植被成为了该项目的主角,随着四季交替呈现出不同的色彩,与北方城市形成鲜明的对比。

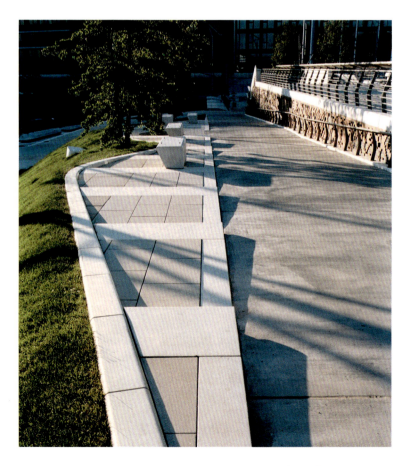

newly planned development, as well as the quayside zones along the harbor basin. Each sector is designed in a way consistent with its intrinsic character. Essential designed features are the special linking of the various public and private surfaces and elevations and links connecting land and water.

The head of the harbor at Grasbrook respond to these uses. Landscape elements are introduced here in a more pronounced fashion. Bomb-damaged areas are planted with weeping willows, and given steps leading down to the water. The head of the Harbor at Grasbrook offers various areas for lingering: two squares set on different levels, an outdoor restaurant, a green hilly landscape with a lawn for reclining under the trees along the water. The landscaping theme is continued all the way to the planting of the marina pontoon.

In the future, Sandtorhafen will be used for historic vessels. The pontoon planned for it constitutes a floating square on the water that constitutes a unity in conjunction with the staircase at Sandtorhafen. The head of harbor at Sandtor is shaped by a spacious, terraced stepped landscape that simultaneously leads down to the water and effects a transition to Sandtor Park. In location, extent, and attractiveness, the surface of the square, resembles a water stage that invites passerby to linger. (Kristine Feirreiss-Aedes)

The architect's intervention is dynamic and flexible: a changing landscape on a human scale, moving partially with the floods, bringing people nearer to the water and its moods. The new urban planning brings the public in a fluid movement from the new housing blocks down to the water, making for everyone's enjoyment a new artificial landscape that is inhabited by natural elements: water and plants. People can find water and trees on every level of the public space.

A big floating platform provides access to small boats, sport boats and ferryboats, as well as leisure areas. Special floating elements provide the presence of greenery and trees at the water level. Water is visible from the borders and through

The open spaces of the western part of Hafencity are central components of the processes of transformation of the former harbor zone south of the historical Speicherstadt (water house district) bordering on the inner city. This area has changed continuously throughout its history, in keeping with various harbor and industrial uses. The alternating ebb and high tides characterize the typical appearance of the port basin.

As protection against storm surges, new mixed used construction surfaces for central inner city functions such as residence, work, commerce, culture, and leisure will be elevated by approx three meters in the course of the development.

In the framework of these processes of transformation from a harbor for ocean going vessels into a inner city district on a human scale, Hafencity Hamburg GmbH announced an open international competition, from which the EMBT emerged as first price winner. As a connective element for the various quarters of western Hafencity, EMBT proposed a lively and multifaceted modulation of the public space.

The planning area encompasses Sandtorhafen, Grasbrookhafen, both harbor heads, and the adjoining squares generated by the

holes, to create a pond like effect. This level is mainly for pedestrians, and will host small cafes thereby creating a relaxed promenade overlooking the water. This level will be flooded only on exceptionally bad weather days, on an average of twice or three times a year. Street level: We propose pedestrian and playing areas also at street level, separating heavy traffic from pedestrians.

A core point of interest in the project is the system of ramps, stairways, and catwalks connecting these different levels. And one of the project's most welcome protagonists is the vegetation; there are many different types and the addition will change the look of the port according to the season of the year, a note of colour and contrasts for the northern city.

项目位置：德国汉堡

客　　户：Hafencity Hamburg GmbH

占地面积：150 000 m²

设计时间：2002 年～ 2012 年

竣工时间：2010 年（部分工程完工）

建筑设计：Benedetta Tagliabue，Miralles Tagliabue EMBT

项目负责人：Karl Unglaub

项目建筑师：Stefan Geenen，Elena Nedelcu

当地建筑公司：WES & Partner Landschaftsarchitekten

所获奖项：2002 年 9 月 Miralles Tagliabue EMBT 与 Thomas Bayrle 合作获竞标一等奖

Location: Hamburg, Germany

Client: Hafencity Hamburg GmbH

Site Size: 150,000 m²

Design Time: 2002~2012

Completed Time: 2010 (several phases completed)

Architects: Benedetta Tagliabue, Miralles Tagliabue EMBT

Project Leader: Karl Unglaub

Project Architects: Stefan Geenen , Elena Nedelcu

Local Architects: WES & Partner Landschaftsarchitekten

Awards: First Prize in a competition in September 2002

Miralles Tagliabue EMBT in collaboration with Thomas Bayrle

世外桃源——莱顿·莱茵公园

Utopia—Leidsche Rijn Park

翻译　李沐菲

DRIE RANDEN

　　莱顿·莱茵公园是 1997 年举行的一个设计大赛的成果。在乌德勒支（荷兰最大的城市之一）西部有一片可容纳 35 000 个住户的新居民区，该公园就坐落在这片居民区内，其设计理念是通过三个边界将公园与周围郊区隔离开来。

　　这三个边界是由一段重新挖掘的莱茵河道、一个生态保护区（长 9000m）和一个围绕公园核心区域（长 4000m）的藤架所组成。公园的核心区域是一片面积为 50 万平方米的绿地，被称为"内院"，里面有森林、水道、行人区、游乐场、林阴大道，为人们创造了一个只有通过大门才能进入的僻静的绿色世外桃源。周围 6m 高的藤架是一面动物志墙，适宜各种各样的动植物生长，从而创造属于自己的小型生态系统。在位置更高的地方还有一个四面环水、藤蔓缭绕的公墓，这也符合了当前荷兰对这些空间利用的需求。公园核心的周围区域是体育场地、公共花园（都市人在其中可以拥有一小块土地，这也就是通常所说的"保持与土地的联系"）和用线性公园连接在一起的其他设施，即"The Jacq. P. Thijsse Ribbon"。登山者、骑自行车和滑旱冰的人们都可以把这个环形公园当做路线，感受到移步异景的情趣。

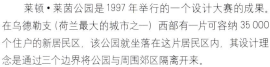

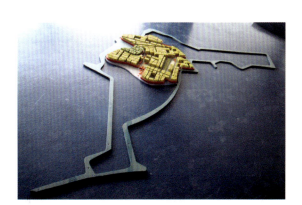

　　该项目平衡了海景房单调的视野，展现出更好的景观。并提供了一个可持续增长的因素。通过有意识地选择生命周期较长的动植物系统，它会比郊区环境发展得更加完善。

The design for Leidsche Rijn Park was the outcome of a design competition held in 1997. The concept for the park, to be constructed in a new residential district of 35,000 homes to the west of Utrecht, one of the Netherlands' largest cities, is based on creating three "edges" that shield the park from its suburban surroundings.

The edges consist of a re-excavated meander of the River Rhine, a nine-kilometer long ecological zone and a four-kilometer long pergola around the core of the park. The core is a 50-hectare green courtyard, called "Binnenhof". It contains woods, watercourses, pedestrian areas, playground and formal avenues which create a secluded green inner world that can be entered only through gateways. The surrounding six meter high pergola is a fauna wall, which will attract all kinds of animals and plants, thus creating its own miniature eco-system. There is also a cemetery lying on higher grounds surrounded by water and the pergola. This is in tune with the current usage of these spaced in the Netherlands. Surrounding the core are sports fields, allotment gardens (communal gardens in which urbanites maintain small plots with small square footage, usually to "maintain a link to the land") and other facilities linked by a linear park of flowery meadows, "The Jacq. P. Thijsse Ribbon". Hikers, cyclists and skaters can move through this circuit park over a track that offers them constantly changing perspectives.

The park is essential to counterbalance the inescapable sea of houses and puts its mono-functional character into perspective. Above all the park will offer a factor of sustainable growth. With its consciously chosen long-term lifecycle it will grow beyond the eternal youth of the suburban environment.

项目位置：荷兰乌得勒支

客　　户：莱顿·莱茵项目办公室

景观设计：West 8

设计时间：1997 年~ 2008 年

施工时间：2005 年~ 2012 年

占地面积：300 万平方米

Location: Utrecht, The Netherlands

Client: Project Bureau Leidsche Rijn

Landscape Team: West 8

Design Dates: 1997~2008

Realization Dates: 2005~2012

Surface Area: 300 ha.

西部花园城市——复兴科伦提巴特居民区

Western Garden City—Kolenkitbuurt Revitalization

翻译　董桂宏

荷兰首都阿姆斯特丹的科伦提巴特居民区建于 1950 年，已经有 60 年的历史了。第二次世界大战结束之后，荷兰人口急速膨胀，急需大量的廉价住房。房地产公司在此时发挥了巨大的作用，他们在 20 年的时间内建造了 200 多万所住宅，标准是每户住宅的建筑面积为 70m²。由于当时地价相对便宜，而建筑材料价格高昂，因此建筑以四五层高的公寓居多，且公寓周围留有大量的公共用地。"西部花园城市"是荷兰国家复兴的原型，由荷兰著名的城市规划师 Cornelis van Eesteren 进行规划设计，而科伦提巴特居民区则是"西部花园城市"的开篇之作，象征着荷兰健全的新型社会福利制度。

但在科伦提巴特 (Kolenkitbuurt) 居民区建成 16 年之后却成了荷兰最贫困的地区，这里大部分的居民都是移民，收入很低、社会流动性大，已成为一个很大的社会问题。房地产公司仍然拥有住房，但他们建房的目标发生了改变，过去他们致力于为低收入人群提供价格低廉的住房，使人人有

所居，而现在他们的目标是提供公众引以为傲的住宅、物业服务和优美的居住环境。根据这一目标，房地产公司不再局限于房产开发，而是探索着投建新的基础设施，他们的服务对象是被荷兰政府称之为"新阿姆斯特丹人"（Nieuwe Amsterdammers）的大批移民。

该项目的总体规划建立在对科伦提巴特居民区的逐步改造的基础之上。在总体规划中规定，所有的住户有权继续在此居住，并且拥有一所新住房。85％ 的住户投票通过了这一总体规划。在实际施工时，房地产公司只有在建好一处新住房后才将原有的住房拆除。迄今为止，总共拆除了 1000 所住房，新建了 1450 所住房，并且对原有的 600 所住房进行了重新装修。新公寓的平均面积达到 125m²，其住房面积相较于阿姆斯特丹城市的平均住房面积相对较大。20 世纪 50 年代所建的"奢华"街道与新住宅非常匹配，街道宽 30m，在街上驾车或步行都非常惬意。新住房旨在提高人们的生活水平，为生活在都市中的人们提供一个舒适的居住环境和适

合邻里交流的空间。设计师对原有的街道也进行了改造，在
靠近住宅的一侧建有 3.5m 宽的行人区，部分地点还建造了
小花园，使其更具生活气息，同时也为一楼的住户保留了适
度的私密空间。各栋住宅之间的公共花园或重建或修复；设
计师还预留了足够的停车用地，保证该地区未来的发展空间。

　　这里原有的居民可以购买新居或者选择租房，房地产公
司与政府还提出了一项新的社会经济方案，即政府为移民提
供成人教育，使他们可以找到更好的工作，并且在这里建立
新的学校和运动设施，保证这里的孩子与其他孩子在同一起
跑线上成长。政府还为在这里投资的企业家提供更多的优惠
政策。

　　现在的科伦提巴特居民区正在经历天覆地的变化。许
多德国建筑事务所参与了该项目的设计，共同打造出该项目
全新的人性化特征。除此之外，后期还将有 15 家建筑事务
所参与其中，该项目预计将于 2016 年完工。

The Kolenkitbuurt in Amsterdam (the Dutch capital) is 60 years. The neighborhood was built in 1950, just after the Second World War. In those days Holland needed a massive amount of affordable housing for a fast growing population. Social housing corporations were the most efficient tools to achieve that goal. These corporations produced in 20 years time more than 2 millions houses according to a national standard: 70 m^2 (gross floor area) for each family. Because space was relatively cheap and building materials expensive, the dominant typology was a 4 or 5 floors flat surrounded by a lot of collective space. The Western Garden Cities, designed by the famous Dutch urban planner Cornelis van Eesteren, is the prototype of a rising nation. The Kolenkitbuurt was the first build part of that: a symbol of the new Welfare State Holland.

Sixty years later the Kolenkitbuurt is officially the poorest and most unwanted neighborhood in The Netherlands. It's a place where mostly immigrants live, barring any possibility of integration; mostly incomes are low; social mobility problematic. Social housing corporations still own the houses, but does

"social" mean these days? In the last decennium these housing corporations redefined their mission statement. An affordable house for people with low income is no longer enough. People should be proud on their neighborhood because it offers great housing, great services and great opportunities for adults and their children. In light of this, housing corporations don't limit their agenda to real estate development anymore, rather attempt to develop a new infrastructure for Nieuwe Amsterdammers (New Amsterdamers), as immigrants are known in the political lingo.

Due to this agenda the master plan is based on a gradual transformation of the Kolenkitbuurt. Eighty-five percent of the inhabitants voted in favor of the plan, in which it was established that all inhabitants have the right to stay and have the right to a new house in a new street. In practice, this means that a building is only demolished once the newer replacement has been completed. In total, 1,000 houses were demolished so far, 1,450 built and 600 renovated. The new apartments have an average size of 125m^2, large floor plan by Dutch – especially Amsterdam's – standards. The original "luxury" street pattern

of the fifties is perfect to adopt new housing. The street width of 30 meters allowed streets where everybody feels at home. Instead of an anonymous urban life, the new-builds are made for people and the people identify highly with their new neighborhood. Supporting this is also the new street profile, with the encroachment zone: a 3.5m wide strip in front of the housing blocks becomes, in some areas, a front garden. It gives the street a more residential outfit and increased privacy to the flats on the ground floor. Common gardens between the housing blocks are redesigned or partly restored. In the new blocks they are elevated by a meter, so that space for parking is gained, tin response to the high demand of parking spots.

The people in the neighborhood can rent of buy a new house. The housing corporations and the municipality simultaneously launched a social and economic program. By being offered adult education older immigrants get new opportunities at better jobs. By developing new schools and sport facilities the children have better chances for a proper head start within society. Entrepreneurs get subsidies when they open new business in the neighborhood.

The neighborhood is now in transformation. A collective of Dutch architects – including HEREN5, VillaNova, FARO,Wingender Hovenier, HMADP, Korth Tielens, Geurts & Schultz, Quist Wintermans, SLA, Thijs Asselbergs and DP6, besides Andries Geerse – was asked to bring a new human dimension to their work. In the coming years another fifteen architectural practices will contribute to the master plan, due to be finished in 2016.

项目位置：荷兰阿姆斯特丹 Bos en Lommer
占地面积：22 000 平方米
项目时间：2005 年 ~ 2009 年
景观设计：Andries Geerse stedenbouwkunduge bv
所获奖项：2008 年 Leeuw van Vlaanderen 国家复兴奖

Location: Bos en Lommer, Amsterdam, The Netherlands

Site Size: 22,000 sqm

Project Dates: 2005 ~ 2009

Landscape Architect: Andries Geerse stedenbouwkunduge bv

Awards: National Renovation Prize 2008 for Leeuw van Vlaanderen

Mario Schetnan

Principal, Architect and Landscape Architect, FASLA
Grupo de Diseno Urbano - GDU (Mexico City, Mexico)
www.gdu.com.mx

In the continent of Luis Barragan and Burle Marx, a new generation of landscape architects grabbed the reigns with gusto.

The Latin American region, a vast and complex geography which borders with the United States of America on the north and generously grows to the south, down to the confines of the Patagonia in Argentina, to the Antarctic pole of Chile and Argentina, is also a vast array of varied cultures and contrasting subcultures, with one main commonality: that of the Spanish and Portuguese languages. However, there is also a sense of identity in the many and varied manifestations of culture and arts: musicians, painters, writers, poets and architects form rich manifestations where "latinos" establish another form of communication, contact and roots.

One of the greatest contradictions in the manifestation of culture in Latin America happens with the profession of landscape architecture. The profession is fairly young in the region and still in the stages of growth and settlement. However, it is astonishing to think that two of the greatest landscape architects of the 20th century were born here, developing and revolutionizing the art of contemporary landscape. They were Luis Barragan in Mexico and Roberto Burle Marx in Brazil.

Burle Marx, the great artist, the committed biologist, plant researcher and collector and the grand executor with its sensuous biomorphic geometry, his amazing variety and plant palette and his masterly manipulation of form at both vast urban scales and small domestic refinement. Much has still to be written by contemporary art historians to acknowledge the fact that this great Brazilian was early in the 1930s (much earlier than Thomas Church, Garret Eckbo or James Rose) creating a contemporary vocabulary and the interest in designing with native plants.

Luis Barragan, who integrated both landscape and architecture in seamless totalities by the poetic manipulation of planar walls, intense use of color and dreamlike fabrications of waterfalls, still mirrors and aqueducts, revolutionized both architecture and landscape in the 1940s and 1950s, transforming functionalism and the International Style into a new "critical regionalism".

Now that both Barragan and Burle Marx have passed away (Barragan in 1986 and Burle Marx in 1994) who are the actors in the following generation in the Latin American region? And what are the new tendencies and manifestations?

Jimena Martignoni, an Argentine critic and writer has recently published the book: "Latinscapes: Landscape as raw material" (by the Spanish publisher GG, Gustavo Gili, 2008). It is an interesting and well-documented book that helps to analyze the region and its performers. I have a general coincidence with the selection of Jimena and have taken it as a basis for the following account of designers and their work.

Starting from the southern most portion of the region, the Chilean landscape architects Juan Grimm, Carlos Martner and German del Sol are three contrasting and interesting designers. Grimm is an artist of plants and acute refinement and has executed mainly private houses and estates of great quality. Martner is an artist, architect and landscape architect, who manifests the region's legacy in public spaces of great interest and power through the intensive use of stone. Del Sol is a young architect and landscape architect who created a true masterpiece in the design of the Puritama hot springs, in Chile.

In Buenos Aires, Argentina, the restoration and recycling of the areas of the old port of Puerto Madero have been interventions of great quality and social and ecological transformation for the city's waterfront. The team of landscape architects has included Alfredo Garay, Margarinos, Vila, Sebastian, Novoa, Joselevich, Cajide and Verdechia.

Rosa Kliass and Fernando Chacel in Brazil: Both Kliass and Chacel worked for periods of time with Burle Marx. Kliass has accomplished great public spaces, in particular wonderful contemporary parks, as with the recent Parque da Juventude in Sao Paulo. Chacel is a master who has incorporated a deep knowledge of ecology and design, creating and restoring mangrove ecosystems with public spaces.

Felipe Uribe in Medellin, Colombia (with Velez and Spera) and Lorenzo Castro and collaborators (Cescas, Leyva and German Samper) in Bogota have recently made new, creative, fresh, urban interventions in urban spaces of great quality and contemporary expression, such as the Barefoot Park in Medellin and the Parque del Agua in Bucaramanga. The city of Bogota has had a truly exemplar transformation through the implementation of parks, linear parks and bike paths, as well as many public spaces that have positively transformed the city. The participation in these plans and projects has been conducted by teams of young, brilliant landscape architects such as Diana Wiesner.

In Mexico, in addition to our office, Grupo de Diseno Urbano, there are several young upcoming designers who have already built public and private spaces of great quality, namely Alejandro Cabeza and Desiree Martinez with the Xochitla Park, north of Mexico City.

However, it is quite impossible and presumptuous to have a fair account of many other old and new players in this vast and rich continent. Please accept this small essay as a brief and incomplete introduction.

建筑师及景观设计师，美国景观设计师协会会员
Grupo de Diseno Urbano — GDU 总负责人（墨西哥墨西哥城）
www.gdu.com.mx

在 Luis Barragan 和 Burle Marx 生活的这片土地上，一批新生的景观设计师开始主宰景观世界。

拉丁美洲北临美国、南部与阿根廷的巴塔哥尼亚相连，一直延伸到智利和阿根廷的最南端，其面积广阔且地形复杂，这里多元的文化和鲜明的亚文化有着一个共同点——那就是都使用西班牙语和葡萄牙语。然而，不同的文化艺术中却蕴含着民族认同感：音乐家、画家、作家、诗人和建筑师竭尽所能地诠释这种民族特性，拉丁美洲人用一种全新的形式进行交流和沟通，这种新方式正慢慢地在拉丁美洲生根。

在拉丁美洲，最大的文化冲突就体现在景观设计行业。景观设计在这里是一个新兴的行业，还处于发展和积累的阶段。但令人惊讶的是 20 世纪最伟大的两位景观设计师却诞生于此——墨西哥的 Luis Barragan 和巴西的 Roberto Burle Marx，他们发展并创新了当代景观艺术。

Burle Marx 是一位伟大的艺术家、住宅资深的生物学家、植物研究学者，善于打造形形色色的生物景观。无论是大都市的建设还是住宅设计，他都能游刃有余地打造出形式多样、颜色各异的植被景观。这位伟大的巴西人在 20 世纪 30 年代（在 Thomas Church、Garret Eckbo 和 James Rose 之前）便开创了用本地植被进行设计的先河，至今仍有很多当代艺术历史学家将其写入书中。

Luis Barragan 善于利用诗情画意的墙壁、强烈的色彩、梦幻般的瀑布、平静的水面以及沟渠将景观与建筑完美地结合在一起，在二十世纪四、五十年代的建筑界和景观设计界都引起了巨大的变革，打破了国际主流的实用主义风格，独具地方特色。

如今，Barragan 和 Burle Marx 两位大师都已辞世（Barragan 于 1986 年逝世，Burle Marx 于 1994 年逝世），谁将成为拉丁美洲景观界的主导呢？未来的发展将呈现出怎样的趋势呢？

阿根廷的作家及评论家 Jimena Martignoni 出版的《拉丁风景：景观设计为本》（由西班牙出版商古斯沃塔·基里于 2008 年出版发行）中用大量的论据生动地分析了当地国情和设计师。我有幸拜读了 Jimena 的这本书，将其视为理解其他拉丁美洲设计师和作品的基础。

拉丁美洲南部的大部分地区都有智利景观设计师 Juan Grimm、Carlos Martner 和 German del Sol 设计的作品的影子。这三位设计师的风格迥异、幽默诙谐：Grimm 是一位植物艺术家，善于精雕细琢，为很多私人住宅和居住区打造了漂亮的景观；Martner 是一位艺术家、建筑师兼景观设计师，他在公共空间的设计上运用大量石块，将当地特色体现得淋漓尽致；German del Sol 是一位年轻的建筑师和景观设计师，他在智利设计的 Puritama 温泉是其杰出的代表作之一。

在对阿根廷布宜诺斯艾利斯的马德罗港口进行重建时，大量的社会资源和生态资源被再利用。景观设计团队中包括 Alfredo Garay、Margarinos、Vila、Sebastian、Novoa、Joselevich、Cajide 和 Verdechia。

巴西的 Rosa Kliass 和 Fernando Chacel 都与 Burle Marx 是同一时代的设计师。Kliass 设计了很多公共场所，尤其擅长设计现代公园，圣保罗的 Parque da Juventude 公园就是出自他手。Chacel 深谙生态学与设计学，他所设计的红树林生态系统经常被应用于各大公共空间。

哥伦比亚麦德林的 Felipe Uribe 同 Lorenzo Castro 和一些波哥大的同事（包括 Cescas、Leyva 与 German Samper）一起为城市美化创造了许多具有质感和现代元素的新景观，如位于麦德林的巴尔福特公园，布卡拉曼加的 Parque del Agua 公园。这些公园、带状花园、自行车道以及很多公共广场彻底改变了波哥大城，而这些设计和工程的主导者是一群像 Diana Wiesner 一样年轻有为的景观设计师。

在墨西哥，除了 Grupo de Diseno Urbano 公司有很多杰出的设计师之外，还有一些后起之秀也有许多不错的公共空间及私人空间设计作品，如 Alejandro Cabeza 和 Desiree Martinez，其中一个作品位于墨西哥北部的 Xochitla 公园。

然而，要想面面俱到地将这片土地上所有的优秀设计师介绍给读者实在是不现实，仅将这篇文章看做是一个简单而不全面的介绍吧。

绿色住宅——海尔加和托马斯·杰特的小屋

Green Residence—Helga and Thomas Jetter Residence

翻译 王玲

该项目位于里约热内卢州安格拉杜斯雷斯市 Porto do Frade 的绿色公寓 12 号、13 号、14 号和 16 号地块上，占地面积为 13 000m²。

项目的亮点之一——木桥，将人们引领到房前，桥下是宽敞如镜的叠层水景，些许"小瀑布"巧妙地顺应地势点缀其间，这些"小瀑布"不仅美化了环境，而且促进了水循环，为观赏鱼和小型水生植物群提供辅助供氧。湖床采用当地石材，周围的群山和葱郁的自然美景倒映在如镜的水面上；草坪巧妙地呼应了附近的高尔夫球场。设计师在 12 号地块上修建门房、马厩和马场。

住宅分为两部分，由一条宽敞的长廊连接在一起。房屋的主立面朝向高尔夫球场，背立面面向一个小院子，院子充分利用房屋的外边缘和当地植被，精巧雅致、郁郁葱葱。为了配合和柔化建筑物体量，设计师运用曲线形的大手笔景观，使之与横平竖直的建筑造型相得益彰，完美融合。

景观设计团队与建筑师通力合作，共同将房屋的社交空间（靠近美食厨房和门廊的区域）打造成一个集多种功能于一体的开阔水池。水池拥有一条 25m 宽的泳道、水上运动区、水疗按摩区、瀑布水景以及从桑拿房直接到达水池和附近平台的通道。花园和水池其他区域的种植槽再现了水池的有机特性，与甲板巧妙地融合在一起。"篝火餐桌"可用于家庭的露天聚会，人们沉浸在葱郁的花园美景中。客房和游戏室位于水池平台的下方。

自然压实的人行道环绕着整个花园，长椅在花园里随处可见；血藤和绿玉藤爬满藤架，争芳斗艳、美不胜收。大部分的原有植物都被保留了下来，特别是本地植物或能够适应当地气候特征的植被，从而有效保护了当地的生态系统。

一条纯净的小溪勾勒出地块的边界，潺潺的流水营造出一种宁静祥和的田园风情。设计师在溪边设计了一块小甲板，人们可以在此放松身心、沉思冥想。

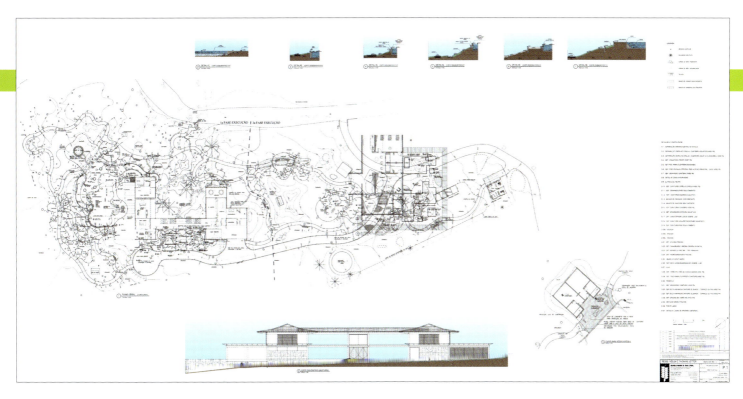

Burle Marx's was responsible for the landscape design and landscape architecture of a total 13,000 m^2 available around this single-family residence located in plots 12, 13, 14 and 16 of Porto do Frade Green condominium, in Angra dos Reis, state of Rio de Janeiro.

A wood bridge – one of the signature elements of the project – provides vehicle access to the plot of land. Underneath, a sizeable, multi-leveled water mirror with several cascades intelligently uses the natural topography of the terrain. The waterfalls serve more than an aesthetical purpose, being used also for water circulation and to provide ancillary oxygenation for the ornamental fish stock and small agglomerations of aquatic plants. The lake bed uses local stone. The water mirror reflects the beauty of the surrounding mountainous, lush natural landscape. Around this area, the landscape architects proposed a lawn that recreates and parallels the adjacent golf course. In lot 12, a house for the property caretaker and a horse stable and arena will be built in the future.

The architecture of the residence is composed of two volumes interconnected by a sizeable covered corridor. The main façade is oriented towards the golf course, while the opposite side faces a small patio with local vegetation and local density, involved by the two volumetric bodies in the outer limits. To complement

and soften the imposing architecture, a large landscape project based mostly on organic, curvy shapes was proposed by Burle Marx, in clear counterpoint with Bernardes & Jacobsen Architects' orthogonal shapes.

Liaising with the architects, the team at Burle Marx, composed by master landscaper Haruyoshi Ono, Isabela Ono and Gustavo Leivas, planned for the social area of the house (close to the gourmet kitchen and the porch) a large, organic-shaped pool, where different functions are assembled: a 25m-wide swimming lane; an area for aquatic games; a hydro-massage tube; a waterfall with a massage bench; direct access from the sauna to both the pool and the contiguous wet deck. Vases reproducing the organic shapes from other areas of the garden and pool involve the deck. Close by, a "table by the fire" was created for intimate family gatherings around an open-air fireplace, surrounded by lush gardens. The guesthouse and the games room are located below the pool deck.

A path for pedestrian circulation surrounds the entire estate, made of natural compacted flooring. Benches furnish multiple niches around the property, and a pergola for climbing plants was planted with the species Mucuna bennettii e Strongylodon macrobotrys, which provide a touch of exuberance in the flowering season. In general, most of the existing vegetation was maintained, specifically native species or species particularly adapted to the regions' characteristics, thus preserving the ecosystem.

Close by, a small creek with pristine water limits the plot. The

running water helps create a serene and bucolic atmosphere. The designers installed a small wood deck by the river, for relaxation and contemplation.

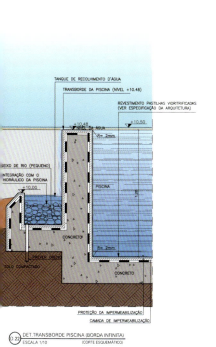

TANQUE DE RECOLHIMENTO D'ÁGUA

TRANSBORDE DA PISCINA (NIVEL +10.48)

REVESTIMENTO PASTILHAS VIDRITRIFICADAS
(VER ESPECIFICAÇÃO DA ARQUITETURA)

+10.48 NIVEL DA ÁGUA
+10.50

R= 2mm.

LEIXO DE RIO (PEQUENO)

INTEGRAÇÃO COM O
HIDRÁULICO DA PISCINA

PISCINA

+10.00

CONCRETO

R= 2mm.

DREVER DRENO

CONCRETO

SOLO COMPACTADO

PROTEÇÃO DA IMPERMEABILIZAÇÃO

CAMADA DE IMPERMEABILIZAÇÃO

D.22 DET.TRANSBORDE PISCINA (BORDA INFINITA)
ESCALA 1/10   (CORTE ESQUEMÁTICO)

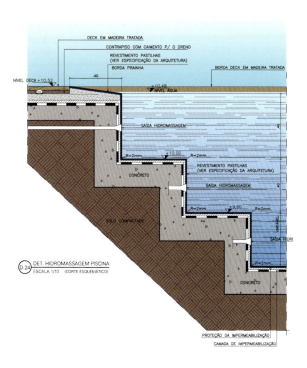

DECK EM MADEIRA TRATADA

CONTRAPISO COM CAIMENTO P/ O DRENO

REVESTIMENTO PASTILHAS
(VER ESPECIFICAÇÃO DA ARQUITETURA)

BORDA PRAINHA

BORDA DECK EM MADEIRA TRATADA

NIVEL DECK +10.52   .40   +10.48
                               NIVEL ÁGUA

SAÍDA HIDROMASSAGEM

+10.00   R= 2mm.
R=2mm.

CONCRETO

REVESTIMENTO PASTILHAS
(VER ESPECIFICAÇÃO DA ARQUITETURA)

SAÍDA HIDROMASSAGEM

SOLO COMPACTADO

+9.60

R=2mm.

SAÍDA HIDRO

R= 2mm.

CONCRETO

D.24 DET. HIDROMASSAGEM PISCINA
ESCALA 1/10   (CORTE ESQUEMÁTICO)

PROTEÇÃO DA IMPERMEABILIZAÇÃO

CAMADA DE IMPERMEABILIZAÇÃO

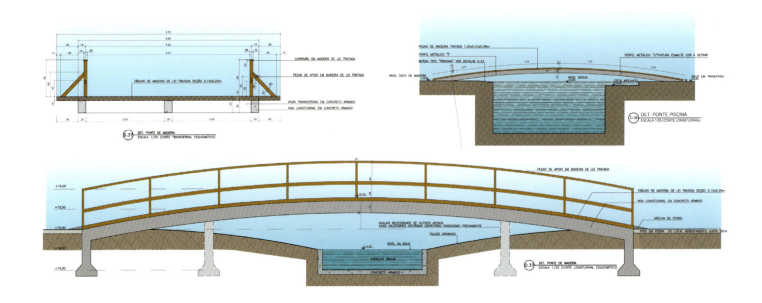

项目位置：巴西里约热内卢州安格拉杜斯雷斯市
客　　户：海加尔和托马斯·杰特
景观面积：13 000 m²
建筑面积：1200 m²
项目时间：2005 年~2007 年
建筑设计：Bernardes & Jacobsen 建筑设计事务所
景观设计：Burle Marx & Cia. Ltda. (Haruyoshi Ono,
　　　　　Isabela Ono, Gustavo Leivas)

Location: Angra dos Reis, Rio de Janeiro, Brazil

Client: Helga and Thomas Jetter

Landscaped Area: 13,000 sqm

Area Constructed in Plot: 1,200 sqm

Project Dates: 2005~2007

Architecture: Bernardes & Jacobsen Arquitetos Associados

Landscape Design: Burle Marx & Cia. Ltda. (Haruyoshi Ono, Isabela
　　　　　Ono, Gustavo Leivas)

乡村别墅——Amanali俱乐部

A Villa in the Country—Amanali Club

翻译　刘建明

PRESA REQUENA

SIMBOLOGÍA
- ▢ VIVIENDA MEDIA
- ▢ VIVIENDA RESIDENCIAL TIP
- ▢ VIVIENDA RESIDENCIAL TIP
- ▢ VIVIENDA RESIDENCIAL TIP
- ▢ VIVIENDA RESIDENCIAL TIP
- ▢ CAMPO DE GOLF
- ▢ SERVICIOS
- ▢ ÁREAS VERDES
- ▢ A. CONSERVACIÓN ECOLÓ
- ▪▪▪ LÍMITE 1a. ETAPA

PLAN MAESTRO ↑

0 100 200 400m

ESCALA GRÁFICA

该项目是一个住宅开发项目，位于墨西哥伊达戈州
Requena 湖（Requena 水库）附近，拥有高品质的 18 洞高尔
夫球场和水上运动设施。

该项目设计旨在将天然的半干旱景观、乡村中心的特质
与传统的广场、高楼和陈列室充分整合在一起，主要的表现
元素是在该场地修筑道路时所挖出的火山岩。植被设计采用
同一原则，将场地上原有的沙漠植被如龙舌兰科的 dasylirion
wheeleri、agave salmiana 和仙人掌科的 Cereus sp.、pipe
cacti 等植被就地保护起来。挡土墙、明亮的屋顶以及围绕着
透明玻璃体的立柱都与房屋的建筑风格形成对比。在负责人
的办公室和销售部的花园中，松散的 tezontle 砾石（西班牙
殖民统治前那一时期的红色火山岩）也与矩形的木质平台形
成对比。穿过由这些巨石组成的石墙，可以看到广阔的湖泊
和山脉景观。

客户要求建筑设计风格简洁，并具有地方特色和现代的
表现语汇，为该项目营造出真实的氛围。

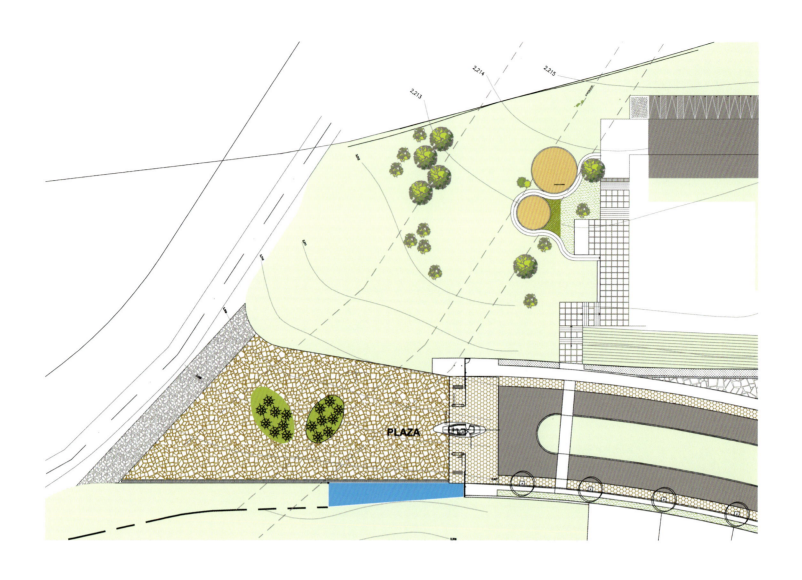

PLAZA

Amanali Country Club and Nautica is a residential development with an 18-hole golf "championship" course and facilities for aquatic sports. It is located by the Lake Requena (Requena Reservoir), in the State of Hidalgo, Mexico. Mario Schjetnan's workshop/Grupo de Diseño Urbano designed the master plan, the landscape architecture and the architecture of the access elements.

The aim was to totally integrate the natural semi-desert landscape, characteristic of the central part of the country, with the formal proposal of plazas, tower-symbol and showroom. The principal expressive element is the volcanic rock that was directly extracted from the site, product of the excavations during the construction of the streets. In the same way, the vegetation based on desert plants like sotol (dasylirion wheeleri), maguey (agave salmiana) and organ (Cereus sp.), pipe cactii were rescued and recovered directly from the site and re-planted. The building's architecture contrasts with the retaining walls and the light roof and thin columns that surround the transparent elements of the glass perimeter. In the garden of the principal office and sales building the loose gravel of "tezontle", a volcanic red rock of Pre-hispanic tradition, makes contrast with the rectangular wood decks. The wide view to the lake and mountainous landscape is framed by massive rock walls.

The clients requested a clean architecture with a very local mood, mixed with a contemporary expression that would establish a real atmosphere for the future construction of the houses in the residential development.

Simbología

■	Zacatule	*Scoenoplectus amer*
■	Tule	*Cyperus bourgaei*
■	Papiro	*Cyperus papeus*
■	Platanillo	*Canna indica*
■	Lantana	*Lantana camara*
■	Rosa laurel	*Nerium oleander*
■	Santolina	*Santolina officinalis*
■	Lavanda	*Lavandula officinalis*
■	Romero	*Rosmarinus oleander*
■	Peniseto	*Pennicetum cetaceum*
■	Pasto kikuyo	*Pennicetum clandesti*

PARQUE LAGOS

0   10   20   30m   ↑

项目位置：墨西哥伊达戈 Tepeji del Río

客　　户：INVERTIERRA，SA

占地面积：294 万平方米

项目时间：2007 年～2009 年（一期）

总体规划、城市设计、景观设计和建筑设计：
GDU Grupo de Diseño Urbano（Arq．Mario Schjetnan）

高尔夫球场设计：亚利桑那州 Schimidt-Curley 设计公司

Location: Tepeji del Río, Hidalgo, Mexico.

Client: INVERTIERRA, SA

Surface Area: 294 ha.

Project Dates: 2007~2009 (phase 1)

Master Plan, Urban Design, Landscape Architecture and

Architecture: GDU Grupo de Diseño Urbano (Arq. Mario
　　　　　　　Schjetnan)

Golf Course Design: Schimidt – Curley Design, Inc. /Arizona

P316

长岛绿城——银杯工作室
Long Island City—Silvercup Studio

P322

多功能户外空间——梅萨艺术中心
Multifunctional Outdoor Space—Mesa Arts Center

P330

国家港口
National Harbor

P336

可持续水资源利用——皇后区植物园新规划
Sustainable Water Resources—New Park Concept for Queens Botanical Garden

P346

制砖厂的华丽转身
The Significant Transformation of a Brick Factory

Janet Rosenberg

Principal, DLitt (Hon.), OALA, FCSLA, ASLA, IFLA, RCA
Janet Rosenberg + Associates Landscape Architects + Urban
Designers (Toronto, Ontario)
www.jrala.ca

City planning in North America is being approached more responsibly today than ever before. We are finally starting to understand the importance of building environmentally sustainable, healthy spaces, partly in response to the growing intensification of urban centers and partly because we are beginning to experience the negative effects of our past actions – and this awareness extends to developers, the general public, and design professionals alike.

As landscape architects, we stand at the centre of this environmental impetus, which means that we have to be very vocal and proactive in terms of educating everyone that open spaces are not just menial additions to projects but they are the necessities and workings of what makes a thriving city. We are speaking up for adequate proportions of parks where people can walk, ride their bicycles, or sit under tree canopies. We are also speaking up for new methods and quality materials that will help to address some of the environmental issues. Our firm, for example, is currently working on a project where we are considering the use of a sub-surface, perforated pipe drainage system in an effort to reduce storm water runoff. What is most surprising is that this simple technique has proven to be very effective but yet it is very low tech and inexpensive.

One of the key challenges cities, such as Toronto, are facing today is the decrease of healthy tree canopies. In re-thinking our cities, we have to ask ourselves: how can we not only increase tree presence but how can we ensure that trees are planted more responsibly to guarantee their longevity? It's not about planting more trees. It's about working collectively with planning departments and developers in finding creative ways whereby tree survival is maximized. This may mean that municipal requirements have to be reassessed and generic classifications have to be adapted accordingly. It's time for us to evolve past cliché approaches to landscapes and find different combinations and juxtapositions of various urban elements that create a new language and a sustainable, innovative focus.

Without a doubt, the globally emerging environmental awareness is resulting in a drastic re-thinking of our cities. In the City of Toronto, for example, we are finally starting to see a shift in emphasis from cars to people in the design and upkeep of public spaces. The fact that, this winter, sidewalks and trails are being plowed by the City (where previously this was the responsibility of individual property owners) is a telling indicator of this shift. I think this is all evidence of a "back-to-basics" mentality in landscape architecture and urban thinking in general, which focuses on people through human-scaled design that is livable, comfortable, and inspiring. This mentality very much involves looking at the successes of public space precedents like Bryant Park in New York or the pedestrian and café culture of European cities, and adapting what works from these models into new designs that fit with the North American lifestyle. Pedestrian-friendly streets, healthy trees, low-tech stormwater management – these are not new ideas, but they have gotten lost in the high-speed life of our cities in past decades. We are finally realizing how important these things are and what a huge impact they can have.

I feel that we are entering into a time where innovative and creative responses to design problems are going to be demanded more than ever, not despite the economic slowdown, but because of it. In my experience, it is during these periods of economic hardship that we are asked to work harder, be more thoughtful in our solutions, and generally put our best work forward. There is an attitude during recession periods where everyone recognizes that resources are limited, so let's design spaces that we can feel proud of. In our office, slow economic periods have historically been periods of growth in the sense of growing our body of work, our ideas, new techniques, and in the sense of building great relationships with our clients. These "growth" periods only work to serve us well in periods of prosperity. Rather than be discouraged by the current conditions, we are looking forward to continuing to advance the cause of great city-building towards new directions.

文学博士（名誉）、OALA、FCSLA、ASLA、IFLA、RCA
Janet Rosenberg 景观设计师与城市规划事务所总负责人
（安大略省多伦多）
www.jrala.ca

目前，北美的城市规划已进入了一个比以往任何一个时代都更加负责的时代。城市中心的密集程度不断升级，由以往建设所带来的负面影响也逐渐显现，在这两者的双重作用下，广大开发商、公众以及专业设计人员终于意识到具有可持续性的、能够健康发展的空间所具有的重要意义。

作为景观设计师，我们是环境学的主要倡导者，这意味着我们必须积极主动地让每个人了解，开放空间并不是项目的附属部分，而是塑造一座繁荣城市所必须的工作。我们一直在争取，希望城市内设有适当面积的公园，可供人们散步、骑脚踏车或者坐在树阴下面小憩。我们也在争取使用创新的方法以及更加高品质的建材，用以协助解决一些环境问题。例如目前正在进行的一项规划，为减少地表径流，正在考虑使用一种多孔管地下排水系统。令人惊喜的是，经证实这一简单的方法行之有效，而且成本低廉。

当今城市，如多伦多，所面临的主要挑战之一便是不断减少的健康树冠。当重新审视我们的城市，不得不这样问自己：怎么能够单纯地增加树木的数量，而不去寻求以一种负责任的态度去种植树木，以保证树木良好的生长并延长其生长周期，这不在于种植的数量，而在于是否能够与规划及开发部门共同合作，找到可以使树木存活率最大化的创新方法。这意味着可能需要重新评估政府的要求，并且根据情况对树种进行分类。现在到了对景观设计旧的模式进行改进的时候了，重新发现城市景观元素的不同组合和方式，用一

种崭新的语言来诠释具有可持续性的、创新的景观设计理念。

全球范围内出现的环境问题无疑引发了人们对所居住的城市的强烈反思。以多伦多市为例，终于看到了这样一种变化，即设计的重点由车辆转变为人和对公共空间的维护。个人认为这些都是景观设计及城市规划理念上"回归本源"的证明，通过更加适宜人居的、舒适的以及令人愉悦的设计来实现以人为本的宗旨。在纽约的布莱恩特公园以及许多欧洲城市的人行步道及咖啡馆，都是以上述理念为主体，我们可以从这些范例中汲取灵感，以创造更加适合北美生活方式的全新设计。人行步道、亲切的街道、茂盛的树木以及低技术含量的排水系统，这些都不是新的方法，但在过去的几十年里，在城市高速发展的过程中已经被人们所遗忘。现在我们终于认识到这些元素的重要性及其对生活的巨大影响。

设计正在进入一个对创新性的需求比以往任何时候都强烈的时代，这并非与经济衰退无关，而是经济衰退加剧了这种需求。但正是因为经济发展进入困境，设计师却需要更加倍努力地工作也因此才能使工作做到最好。经济衰退让每个人都认识到了资源的有限性，让我们设计让人自豪的空间景观吧。从历史发展的角度来讲，经济低速发展期正是行业的高速成长期，包括设计工作主体、理念和新技术；也是与客户发展良好合作关系的成长时期，为经济复苏后的发展打下了良好的基础。

沙漠之花——Safari Drive

A Flower in the Desert—Safari Drive

翻译　李沐菲

　　该项目坐落于美国亚利桑那州的斯科茨代尔市，因著名的 Safari 酒店而得名，周围是一片生机盎然的工作与生活社区。该项目地处亚利桑那州运河北岸，坐拥众多零售店及写字楼。该项目占地面积为 36 422m²，其设计主旨是建造一系列的庭院和露台，用以创造更多户外活动的机会，并通过绵延起伏的群山景观来突显沙漠气候的特色。

　　该项目的主体结构采用跨越式设计，其主要特色是高处的不锈钢植栽容器以及地下车库（用以满足该区域大部分的停车需求），这一开放式空间规划营造了一个强调户外生活模式的度假型社区。

　　具有现代风格的零售店以及写字楼均采用了视野开阔的大窗设计，将周围大胆而抽象的景观尽收眼底；坚固的花岗岩道路直接通往亚利桑那州运河北岸新兴的周边设施；浓密的草坪配备有防火和防水设施，成为户外活动空间；郁郁葱葱的椰枣林中间设有两个泳池，展现出浓厚的度假氛围。

　　该项目东部的边缘正对着亚利桑那运河，在斯科茨代尔市政府以及文化局的共同支持下，这一地区正被改建为斯科茨代尔当代艺术体验馆（SCAPE）。作为一个整体景观项目，SCAPE 将成为一座动态的线性公园，也成为众多当代艺术家及设计师作品的展示场馆。

Located in the heart of Scottsdale, Arizona, Safari Drive was inspired by the historic Safari Hotel, which once neighbored this vibrant live-work community. Safari Drive sits on the north bank of the Arizona Canal and is surrounded by retail and commercial office space. The design goal for this 9-acre retail and condominium site was to create a series of courtyards and terraces to promote outdoor living opportunities and celebrate our desert climate by taking advantage of sprawling mountain views. The majority of this project is over-structure, featuring raised steel planters and a below-grade garage structure to accommodate the majority of the areas' parking needs. This open site plan creates a resort-like community with a strong emphasis on outdoor living opportunities.

The contemporary architecture of the retail and office space incorporates vast windows which open onto the grand views, complimented by the bold abstract forms of the surrounding landscape. Stabilized granite paths provide direct pedestrian links to the revitalized north bank of the Arizona Canal frontage and surrounding amenities. Intimate lawn terraces with fire features and water features serve as outdoor living rooms. Two pool terraces nestled into date palm groves with lush planting provide a resort-like atmosphere.

The eastern edge of the property fronts a portion of the Arizona Canal which is concurrently being transformed into the Scottsdale Contemporary Art Place Experience (SCAPE), in partnership with the City of Scottsdale and the Scottsdale Cultural Council. As a whole, SCAPE will provide a dynamic linear park designed to showcase the work of a variety of internationally known artists and designers.

项目位置：美国亚利桑那州斯科茨代尔市
占地面积：36 421m²
景观设计：Floor 公司（JJR I Floor）
建筑设计：Miller Hull Partnership
预　　算：100 万美圆（一期）；75 万美圆（二期）
建成时间：2008 年（一期）；2009 年（二期）

Location: Scottsdale, Arizona, USA

Surface Area: 9 acres

Landscape Architecture: Floor Associates (JJR I Floor)

Architecture: Miller Hull Partnership

Budget: 1 million (phase one); $750,000 (phase two)

Completed Time: 2004~2008 (phase one);

2005~2009 (phase two)

城市的剧场——内森·菲利普斯广场改造

A Theatre for the City—Nathan Phillips Square Revitalization

翻译　王玲

大约公元前 600 年～ 350 年，希腊雅典在市内两个地点开启民主实践活动：城市广场和剧院——城市广场让人们走出自己的小圈子，去关注城市里其他人的存在和需求；剧场则有助于人们在做决定时集中他们的注意力。

——理查德·桑内特（Richard Sennett）《民主的空间》

威里欧·若威尔（Viljo Revell）最初设计的内森·菲利普斯广场是一个市政广场或古希腊式的集市。在多伦多市政厅设计的竞标入围者中，只有威里欧的设计将一片宽敞的空地与市政厅标志性的会议大厅并置。由于市民没有直接参与决策，公共广场就面临着要满足广场和剧院这两种功能的压力。威里欧最初的设计是通过独特的建筑立面来诠释广场和剧院这种双重功能性的，但随着广场边缘的逐渐升高，设计重点便转到了室内。尽管古代雅典广场上的拱廊设计充当了开阔的广场和附近商铺会馆等私密空间的分界线，但是在内森·菲利普斯广场上，这种空间往往被忽视。虽然一些市民对广场的空旷感觉不适，但设计师坚持认为这种开阔设计正是民主力量的来源。广场中心的树木被移走，这里便成为了一个可以举办各种聚会和活动的纯粹空间。

从建筑学的角度来看，广场水池西面重新改建的凉亭对广场的布局十分重要，并与人行道相连接。广场上点缀着灯饰、水景和石板凳，充满生机与活力。环绕停车场的楼梯利用玻璃材质重新建造，并安装嵌入式灯以满足广场的夜间照明所需。该广场可以说是开启多伦多新城区活力中心的一把新钥匙。

绿树掩映下的房屋紧密环绕在广场四周，不同设计元素的组合形成了不同的景观体验。各种树木、地被植物和屋顶植物不经意间勾勒出一幅田园般的风景画：每一片林木区都是简单随性的设计，通过 9 种植被的不同组合，逐渐过渡形

成适应其周边环境的景观设计。处于规划阶段的小树林里满是各种本地植被，它们枝繁叶茂，树影婆娑，适应着城市的生活环境。冬天，它们叶落茎枯，形态各异；秋天则层林尽染，色彩斑斓。

市政厅的裙楼不仅是人行道的一部分，也是广场周围花园中最大且最幽僻的地方。它位于开阔的广场中心和场地周

围绿树掩映下的房屋之间，生动地勾勒出公共广场的线条。作为一个大露台，也充当了城市剧院的夹层；其下面是多伦多闻名于世的地下通道。在恶劣天气下，地下通道是连接市中心与市政厅的一条重要的人行通道。为了增强当前交通动线的效率，主楼梯被改建成旋转楼梯，行人从皇后大街上的地下通道出来便可以看到内森·菲利普斯广场和市政厅的入口。

和平花园彻底改变了设计师对广场的设想。当前的空间设计打破了原本的开阔性，剧院成为其折中选择。开阔的和平花园位于人行道和奥斯古大厅之间的广场西侧，环绕一个狭长的黑色花岗岩浅水池而建。浅水池里的水非常浅，人们可以在上面自由行走而不用担心弄湿鞋。浅水池中间的一条花岗岩石桥将和平花园与广场连接在一起。一条狭长的坡道从石桥向南一直延伸到黑色花岗岩墙。人们沿着坡道缓缓上行，浅水池尽收眼底；随着坡道穿过墙面，将和平花园和人行道连接在一起。

设计团队将内森·菲利普斯广场视为城市的活力中心。这里是展示构成多伦多复杂文化的根本差异性的理想场所。作为一处展示文化活动的非营利场所，广场为社会各界提供了一个免费的可供展演的真正民主空间。设有慢行限制的广场周边地区也是开展个人或小型社会活动的场所，这里是远离人群、聆听"少数话语"的理想私密空间。

Athens, from roughly 600 to 350 BC, located its democratic practices in two places in the city, the town square and the theatre. The Square stimulated citizens to step outside their own concerns and take note of the presence and needs of other people in the city. The architecture of the theatre helped citizens to focus their attention and concentrate when engaged in decision making.

—The Spaces of Democracy by Richard Sennett

Viljo Revell's design for Nathan Phillips Square was originally conceived as a civic square, or agora, at the scale of the city. Of the finalists in the Toronto City Hall competition only his project juxtaposed an enormous void with the City Hall's iconic council chamber. Because citizens do not participate directly in decision-making, pressure is put on the public square to serve the democratic functions of both square and theatre. Viljo Revell's original design for the square claims this dual function through its unique section, with its raised edges that focus views towards the interior. While the porch, or stoa, in the old Athenian agora provided a threshold between the openness of the square and the intimate spaces of shops and assembly halls adjacent to it, at Nathan Phillips Square this set of spaces has always been missing. While some Torontonians feel uneasy about the emptiness of Nathan Phillips Square, the designers assert its openness as the source of its democratic power. By removing the trees from its center, the square is left as a space of pure potential for varied interactions and events.

The renovated pavilion west of the pool will bundle essential support for occupation of the square under a new overlook architecturally connected to the raised walkway. The surface of the square will be animated by light, water, and stone slab seating placed on the square's grid. The stair enclosures from the parking will be rebuilt in glass with recessed lighting to create beacons on the square at night. Opening the square is the key to unlocking its potential as the active center of the enlarged City of Toronto.

An intimate perimeter will be formed by tree-canopied rooms, whose varied ambience will be constructed through the juxtaposition of different elements. Trees, ground covers

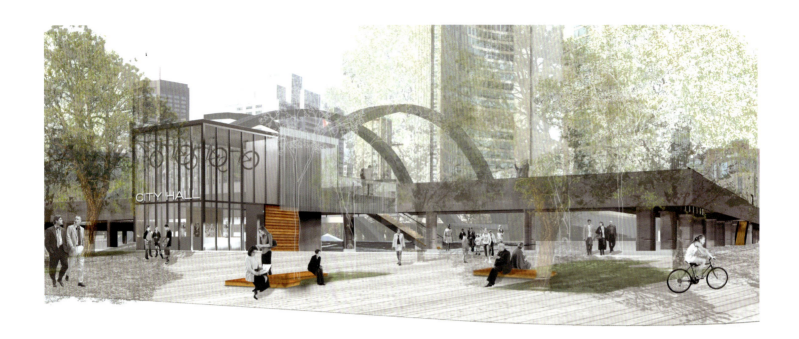

and roof plantings are all arranged in varied informal mosaic like field: each treed area is an informally arranged mix of trees, which slowly transforms using different combinations of 9 species to suit the different exposure conditions. The proposed tree groves combine natives with one hybrid, strongly emphasizing shade, urban tolerance, variety of shape for winter sculptural interest, and fall colour.

The podium of the city hall building is at once part of the raised walkway system and the largest and most secluded of the square's perimeter gardens. It acts as the threshold between the open center and the treed green rooms at the perimeter of the site. This liminal space acts as the defining line of the public square. As a balcony, it introduces a mezzanine section completing the theatre of the city. Down below, Toronto's world-famous underground walkway, the PATH, is an important foul weather pedestrian connection from downtown to City Hall. In order to improve the current route the main stairs were rotated so that Nathan Phillips Square and the entrance to City Hall present themselves to the pedestrian emerging from the PATH on Queen Street.

The Peace Garden has substantially altered Viljo Revell's vision for the square. The current space obstructs the openness of the square compromising its potential as a theatre. The project proposes an enlarged peace garden to the west of the square, between the raised walkway and Osgoode Hall. It will be organized around a long black granite pool, with water so shallow that people can walk through the pool without getting their shoes wet. At the center of the composition a granite bridge spans the pool connecting the peace garden to the interior of the square. Running south from the bridge, a long ramp hugs the black granite wall, overlooking the pool as it rises, puncturing the wall and bridging from Peace garden to raised walkway.

The designing team sees Nathan Phillips Square as the dynamic heart of the city. It is empty precisely to accommodate the radical differences that constitute the complex cultures of Toronto. As a non-profit infrastructure for programmed cultural

events, the square provides a truly democratic venue for communities without economic means to secure expensive performance space. This slower perimeter zone is also a space for individual and smaller social interactions, offering intimate spaces for the emergence of minority discourses outside the exposed public spaces of the central square.

位　　置：加拿大安大略省多伦多市
开工时间：2007 年
景观设计：PLANT 建筑设计事务所（主要设计团队）

Location: Toronto, Ontario, Canada
Project Start Date: 2007
Landscape Architects: PLANT Architect Inc. (head team)

亲水设计——振兴湖滨区

Accessing the Water—Water's Edge Revitalization

翻译　王玲

　　湖滨中心是多伦多最主要的文化中心和休闲区之一，也是多伦多中央滨水区的重要组成部分。自 2005 年 6 月正式开放以来，滨水区振兴的规划标志着大滨水区一体化发展迈出第一步。

　　历经 30 多年的岁月洗礼，湖滨区的公共空间已显现出倦容与沧桑感。虽然游客没有意识到透过酒店窗户观赏到的景致别有一番风情，但是与酿酒区等新区相比，这里早已黯然失色。然而，湖滨中心和多伦多滨水振兴公司（现称为多伦多滨水管理处）发现了振兴城市文化遗产的价值，从而促进了滨水区发展的良机。

　　建筑师、景观设计师和工程师组成的设计团队提出了一个全面的滨水城市设计方案，采用"三管齐下"的振兴策略——即重塑滨水区形象，强化通达性和舒适性；改造建筑、景观和公共空间；促进各种活动的动态交融。

　　该项目的一期工程位于约克码头和锡姆科大街之间。已完工的项目中包括一条新的滨水景观步道，步道两边设有专

门定制的节能灯、花岗岩长凳和系船柱，堤岸景色优美，周边有林阴小径和一条悬浮木栈道。码头与木栈道垂直，并延伸至多伦多港湾。

　　引人瞩目的振兴规划还包括一个吸引游客和当地居民的公共空间——即将滨水区的各种表演场所和手工艺品工作室聚合在一起的"中央文化滨水区"，这样不仅创造了一个可以俯瞰美景的宽敞的公共步道，还可以为手工艺展示、街头表演等非正式活动提供了灵活自由的空间。滨水区振兴规划开创了一门新的城市设计语言，无论东部的港口还是西部的海滩，这门新的城市语言在许多新项目中都被成功演绎。

亲水设计

　　设计师将混凝土挡土墙、系船柱、瞭望台和楼梯全部拆除，有效地消除了游客与滨水美景之间的屏障。一系列分布在防波堤旁的花岗岩压顶石充当了公共坐席，人们在此可以尽赏水天一色的美景。坡道和平缓的高差设计取代了原来的

阶梯，从而使交通更加便捷；另外，还有一条悬浮木栈道伸向水面。

文化关联

　　绵延的步道将场地上的四个主要的文化场所连接在一起——约克码头中心、发电厂、Enwave 剧院和 1927 年建的码头仓库和首演舞蹈剧院以及各种商店、餐厅。原来的楼梯、坡道和挡土墙被平坦的铺装所取代，而步道则成为建筑间的连接廊道。

城市设计语言

　　设计师开创了一种由材料和设计元素组成的生动语言，彰显出现代海岸线这种高度城市化的特质。花岗岩、混凝土和钢材等建筑材料与湖滨区的工业历史相得益彰；连绵的屋顶轮廓线、肋状屋顶结构和入口处凉亭的钢面与进出港湾的货轮交相辉映，夜幕下灯光璀璨，这一切仿佛约克码头入口的一面标志性的旗帜。

滨水分层式设计

　　振兴规划的景观元素分为不同的层面，它们体现并强化出湖滨区的地域风貌。游客来到这里，沿着步道、林阴小径、堤岸、长凳、木栈道、栏杆以及系船柱，最终到达风景旖旎的水边。这些景观元素勾绘出场地的独特气质，不仅强化场地的西边界锡姆科大街和东边界约克码头，同时也增强了约克码头和约克大街之间历史性的南北贯通。

丰富的公共空间

　　沿着西步道和湖畔的开放空间是音乐表演、手工艺品展示和街头表演的临时舞台。设计巧妙地将工业设施隐藏了起来，削弱了场地的工业化印迹，营造出更多井然有序的公共空间。设计团队提出将公共空间向北延伸，使之取代码头仓库旁的停车场，多伦多滨水管理处（多伦多滨水振兴公司前身）正在开展这一规划的可行性研究和方案设计工作。

可持续性设计

　　由于多伦多位于北美候鸟迁徙的主要路线上，因此约克码头的照明系统是在不危及迁徙候鸟的情况下保障城市夜间安全。新的 LED 照明系统耗电量仅为传统公共照明系统的 1/10，它也是迄今为止加拿大最大规模的 LED 照明系统。远离约克码头的地方，设计师利用 24 株枯树、2500 吨的毛石混凝土和石块修建了一座 370 m² 的水下鱼类栖息地。

The Harbourfront Centre is one of the city's major cultural and recreational venues, and a focal point of Toronto's Central Waterfront. Officially opened in June 2005, the Water's Edge Revitalization Project marked the first step in the creation of a continuous and accessible water's edge that will eventually stretch across the entire Waterfront.

After some 30 years of operation, Harbourfront's public spaces were worn and weathered. Visitors were unaware that the site visible from their hotel windows contained venues worth their attention, and the Centre had lost profile among city residents at a time when it needed to compete with new venues such as the Distillery District. Harbourfront Centre and the Toronto Waterfront Revitalization Corporation (now Waterfront Toronto), saw an opportunity to revitalize a major City cultural asset and to further the ongoing process of waterfront revitalization.

A design team of architects, landscape architects and engineers developed a comprehensive urban design plan for the water's edge that employs a three-pronged renewal strategy: re-shape the water's edge to increase access and amenity; reconfigure buildings, landscapes and open spaces; and promote a dynamic mixture of public uses, attractions and activities.

The design team undertook Phase 1 of the Revitalization along the water's edge between York Quay and the Simcoe Street Slip. The completed project included a new landscaped waterfront promenade furnished with custom-designed energy-efficient lighting, granite benches and bollards, a landscaped berm and allées of new trees, and a floating boardwalk with finger piers projecting out into Toronto Harbour.

This highly visible Revitalization has created an enhanced public space that draws visitors and residents alike. It unites Harbourfront's various performance spaces and craft studios into a central "Culture Port" on the waterfront, creates a spacious public promenade directly overlooking the water, and provides flexible, un-programmed public spaces for craft shows, street performance and other informal uses. The Water's Edge Revitalization project has also provided a language of urban design elements that is being applied to new developments from the Eastern Portlands to the Western Beaches.

Accessing the Water

The designers eliminated barriers between visitors and the water by removing concrete retaining walls and lookouts, bollards and stairs. A continuous granite capstone along the seawall at the promenade edge provides public seating without obstructing views of the water. Ramps and smooth changes in grade replace pre-existing stairs, creating a universally accessible site. A floating boardwalk projects over the water.

Culture Connections

The continuous promenade links the site's four significant cultural facilities: the York Quay Centre, the Power Plant, the Enwave Theatre and the 1927 Terminal Warehouse, home of the Premier Dance Theatre and a wide variety of stores and restaurants. Pre-existing stairs, ramps and retaining walls have been replaced by level paving, and the promenade reads as a connecting hallway between buildings.

Urban Design Language

The designers have created a language of materials and design elements that emphasize the highly urban character of the modern shoreline. Building and paving materials – granite, concrete and steel – echo the industrial history of the Harbour. The sweeping roofline, ribbed roof structure and steel skin of the Entry Pavillion echo the lines of the cargo ships that still ply the harbour. Illuminated at night, it serves as a beacon marking the entrance to York Quay.

Layered Design at Water's Edge

The elements of the Revitalization project read as a series of layers that define and reinforce Harbourfront as a precinct. As visitors approach the site, they experience in succession the Promenade, the buffering allées and berms, benches, boardwalk, railings and bollards, arriving finally at the water's edge itself. These elements delineate the site, reinforcing its western boundary at the Simcoe Street Slip and eastern boundary at York Quay, and highlight the historic north-south connection between York Quay and York Street.

Enriching Public Space

Open spaces along the west promenade and lakefront accommodate temporary stages for music festivals, craft exhibitions and street performers. Concealing site services "de-industrializes" the site and creates more organized and significant public spaces. The design team proposed the northward extension of public space to replace an at-grade parking lot beside the Terminal Warehouse. Waterfront Toronto (the former TWRC) is now conducting a feasibility study to develop options based on this proposal.

Sustainability

Toronto is directly in the path of one of North America's major avian migratory routes. Lighting on York Quay addresses urban concerns for nighttime safety and security, without creating dangerous distraction for migrating birds. The new LED lighting system uses only 1/10th the energy of a conventional public lighting system, and is the largest such installation in Canada to date. A new 370 m^2 underwater fish habitat has been constructed off the York Quay using 24 dead trees and 2,500 tonnes of concrete rubble and stone.

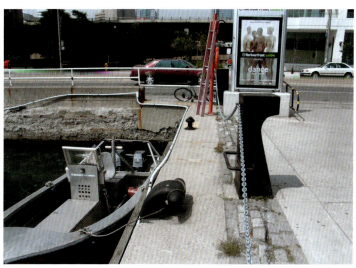

项目位置：加拿大安大略省多伦多市
客　　户：湖滨中心／多伦多滨水振兴公司（多伦多滨水管理处）
预　　算：约 1030 万美圆
建成时间：2005 年
建筑及城市设计：architectsAlliance
景观设计：Envision—The Hough Group
所获奖项：2007 年多伦多城市设计优秀奖

Location: Toronto, Ontario, Canada

Client: Harbourfront Centre / Toronto Waterfront Revitalization (Waterfront TORONTO)

Budget: approx. US$10.3 million

Completed Time: 2005

Architecture and Urban Design (lead): architectsAlliance

Landscape Architecture: Envision – The Hough Group

Awards: City of Toronto Urban Design Award of Excellence, 2007

沙漠绿洲——亚利桑那州立大学生物设计研究所

An Oasis in the Desert—The Biodesign Institute at Arizona State University

翻译　王玲

为了打造一处收集雨水兼城市野生动植物栖息地的生态海绵，坦恩·艾克景观设计事务所设计了一个示范性的户外沙漠广场。该广场不仅有效地满足了生物研究所研究和教育的要求，而且还成为了一处舒适宜人的聚会场所。亚利桑那州立大学生物设计研究所是亚利桑那州立大学生物学、生物工程学和纳米科学系统领域最先进的机构。作为校园东门和生物研究所前的广场区域为在纷繁的功能性公共空间中发展可持续性设计提供了良机；保留区和旧停车场等剩余空间则被改造成一个彼此相连、生机盎然的索诺兰沙漠区，这里主要展示了一些自然环境中的生物群落。设计团队的工作包括总体规划、校园新东门的设计以及用于讲座、教学和社交聚会的户外空间的设计。

环境保护与提升

该场地上原有一个由干枯河流中的石桩组成的草地滞留池和一个沥青停车场。设计师计划改造滞留池，使其更好地服务于该场地和校园。设计师将当前的场地彻底改造成为一个重建的沙漠保留区和一个种植了豆科灌木的绿阴广场。由鹅卵石铺设的渗透池经过自然历程具有了传统水景的特色，结合丰富的自然材料产生同样轻松愉悦的感官享受，同时渗透池作为人们喝咖啡的休闲空间，亦成为了广场的中心元素。广场采用可渗透风化花岗岩铺面，不仅增强了沙漠灌木丛的原生态，而且在减少硬质铺面数量的基础上增强了雨水的自然渗透力。自行车道选用与特勒斯街（Terrace Street）相同

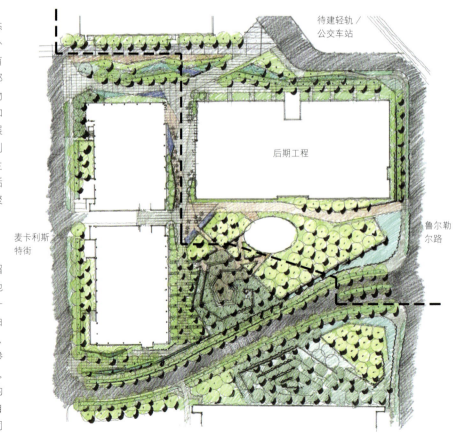

的混凝土材料，这样不仅可以有效地区分车道，而且有窄化路面的视觉效果。与宽阔的沥青路面相比，这种路面不仅在视觉上更加舒适，更有助于减慢车速，为行人和骑自行车的人创造更加舒适宜人的环境。

与自然环境的交融

植被的选择也呼应了沙漠的不同层面。沿建筑的柱基种植着假紫荆属植物，与保留区内的豆科灌木和棱角分明的现浇混凝土形成鲜明对比。渗透广场由泥土构成，周围是用于休息的台阶和充满生机的"生态海绵"——一个沙漠滨水花园，不仅可以收集新建筑屋顶上的雨水，而且在本地豆科灌木斑驳的树阴下创建了城市野生动植物栖息地。花园初期的灌溉是由雨水完成，待建成后将利用另一建筑中空调系统的冷凝水进行灌溉，从而有效地减少饮用水的消耗。

与人工环境的交融

在该项目中，花园与建筑不仅和谐共存，而且彼此影响。建筑收集到的雨水可用于浇灌花园，花园则可以作为舒适宜人的户外聚会空间，将生物研究所的工作带到户外。抬高的建筑与下沉广场巧妙结合，不仅彰显了建筑的气质，而且也体现出滨水景观的闲情逸致。植被过渡带将人行道和建筑立面分割开来，使得建筑入口更加突出。植有树木的上层广场不仅是进入建筑北侧的过渡区，同时也将空间更加紧密地连接在一起。为了迎合柔和的表面效果，设计师尽量减少硬质

景观的使用，因为这种柔和的表面有助于将沙漠美感更好地融入到城市环境中。

自然资源的保护

在迷宫一样的圆形花园周围是场地上最具特色的景致——生态湿地。雨水和屋顶排水被收集到生态湿地中，并仿佛一块海绵般对缓缓流入低处豆科灌木林的雨水进行处理后，才最终汇入到收集系统中。在不久的将来，花园将与场地附近的一座建筑连接起来，并完全利用雨水和空调冷凝水进行浇灌，这也就意味着无需任何自来水便可实现对绿阴覆盖的沙漠花园进行维护。将原来的草场保留区改造成一个沙漠栖息地，这样设计在很大程度上减少了场地对灌溉用水的需求。虽然场地上全部采用沥青铺面，但该项目并未破坏沙漠的原生态，在节水设计和本地植物的营造中体现出索诺兰沙漠的城市片断。

促进环保的认识

由于沙漠植物景观被分成许多不同的区域，因此人们在这里可以亲眼目睹到假紫荆属植物与沙漠流水旁的其他植被交相辉映、和谐共生。作为通往大学和生物研究所的大门，场地每天需要接待大量的人群，而经过此处的人都会被生态湿地花园中清爽的阴凉和迷人的水景所吸引。规划中的会议中心也将建于花园中，来自世界各地的人们将会在纷繁的城市生活中感受到别样的沙漠风情。

In designing a bio-sponge that harvests rainwater and provides urban wildlife habitat, Ten Eyck created a didactic outdoor desert plaza that complements the cutting-edge research and education taking place at the Biodesign Institute and also serves as a comfortable gathering space. The Arizona State University Biodesign Institute is a state-of-the-art facility in the fields of Biology, Bioengineering and Nanosystems/sciences for Arizona State University (ASU). The setting as both an eastern gateway to the campus and the foreground of the Biodesign Institute provided a perfect opportunity to promote sustainable practices in a busy, functional public space. Thus, in Ten Eyck's project the leftover spaces – a retention area and old parking lot – become a cohesive, thriving Sonoran Desert habitat showcasing several biomes in naturalistic settings. The team's work included landscape architectural master planning, design of a new east entry to the campus and the design of outdoor spaces for lectures, learning and social gatherings.

Preservation or Enhancement of Environmental Quality
This site previously included a grass retention basin comprised of barren river rock piles and an asphalt parking lot. The decision was made to include the "renovation" of the retention basin in the scope of work to make it more of an asset to the site and campus. From those conditions, the current site has undergone a complete transformation to include a reworked desert retention area and a shaded amphitheater tucked in a cool mesquite bosque. A river rock seep basin puts a spin on the typical water feature by using a natural process to provide the same soothing effect with more natural materials, and acts as the focal point of the plaza with space for café tables and chairs. The desert plaza features a permeable decomposed granite floor that emphasizes the feeling of a natural desert bosque, decreases the amount of hard, paved surfaces and allows for natural infiltration of rainwater. We replaced the material of the bike lanes with concrete on Terrace Street to distinguish the lanes and narrow the visual appearance of the roadway. Not only will this be easier on the eyes than a wide expanse of asphalt, the narrowed roadway will help to reduce speeds and create a more comfortable atmosphere for pedestrians and bicyclists.

Integration and Compatibility with the Natural Environment
The planting scheme was designed to evoke different levels of a natural desert. The raised plinth along the building, 5' above grade, is planted with the paloverde creosote habitat, which is contrasted with the mesquite bosques and angular cast-in-place concrete in the retention areas. The permeable desert plaza is sculpted out of the earth with seat walls and is surrounded with a living, breathing "bio-sponge"—a desert riparian garden that harvests rainwater from the roof of the new building and creates urban wildlife habitat under the filtered shade of native mesquite trees. In the initial phase, the gardens are irrigated with an irrigation system supplemented with rainwater. The gardens are set up, however, to be watered in the future with condensate from the air conditioning system of another building in the complex, reducing the amount of potable water needed for irrigation.

Integration and Compatibility with the Manmade Environment
This is a case in which the gardens and architecture do more than

just exist together; they each make it possible for the other to thrive. The buildings supply the gardens with collected rainwater runoff and the landscape provides comfortable outdoor gathering spaces that can help bring the work of the Biodesign Institute outdoors. The raised building is complemented by a sunken garden in a manner that showcases both the elevated architecture and the recreated wash featured in the riparian zone. A planted buffer separates the sidewalk and the façade, providing more emphasis to the building entrances. An "upper" plaza with trees provides a transition to the building on the north side and will knit the space together. Hardscape materials were minimized in favor of softer surfaces that give help bring the beauty of the desert into the urban setting.

Conservation of Natural Resources

The labyrinth-like amphitheater garden is embraced by the site's most striking feature: the bio-swale. Storm water and roof drain run-off is directed into the bio-swale and acting like a sponge, the bio-swale allows native vegetation to treat the water as it slowly meanders and descends into the lower mesquite bosque before entering the retention system. A future connection to a nearby building on the site will allow the garden to be irrigated solely with water from rainfall runoff and air conditioner condensate,

meaning no potable water will be needed to maintain this shaded desert garden. By renovating the old turf retention area to a desert habitat, we drastically reduced the need for irrigation water on the site. All construction occurred on land that had been covered with asphalt – no pristine desert was degraded to build this project, and we used all low water use and native plants to create our urban slice of the Sonoran Desert.

Contribution to Environmental Awareness

Because the planting showcases several levels of desert planting zones, future visitors to the site will have the opportunity to enjoy several natural plant communities as they coexist in the Sonoran Desert. This is quite a different approach to typical landscape planting that includes plants from around the world. Visitors will see firsthand the subtle comparison of the paloverde creosote association with the variety of plants that follows running water through the desert. The site's capacity as a gateway to the university and its proximity to the Biodesign Institute will undoubtedly bring a large number of people each day who will be drawn to the bio-swale garden for its cool shade and its fascinating use of water. A planned future conference center nestled into the garden will attract international visitors that will have the chance to experience the Sonoran Desert in a busy urban area.

项目位置：美国亚利桑那州坦佩市

项目时间：2002 年 ~ 2006 年（一期和二期）

客　　户：亚利桑那州立大学

建 筑 师：古尔德 • 埃文斯

景观设计：坦恩 • 艾克景观设计事务所

所获奖项：美国景观设计师协会 2009 年荣誉奖

　　　　　能源与环境设计先锋奖（LEED）白金级认证——二期

　　　　　2005 年 Valley Forward 环境优秀奖 /

　　　　　大型社区发展奖——Crescordia 奖

Location: Tempe, Arizona (USA)

Project Date (phases 1 and 2 only): 2002 ~ 2006

Client: Arizona State University

Architect: Gould Evans

Landscape Architect: Ten Eyck Landscape Architecture

Awards: ASLA 2009 Honor Award

　　　　Phase 2 achieved LEED Platinum;

　　　　2005 Valley Forward Environmental Excellence Awards Program/

　　　　Large Scale and Community Development Category —

　　　　Crescordia Award

巴雷尔·韦尔豪斯公园

Barrel Warehouse Park

翻译　刘建明

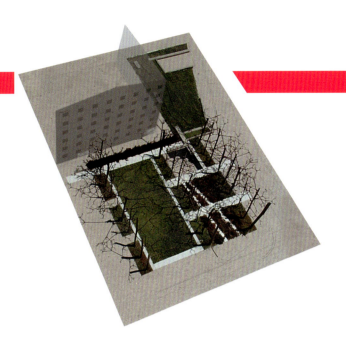

该项目位于沃特卢市新中心区的核心地段，靠近历史上有名的西格拉姆酿酒厂以及新近被合二为一的两个仓库。该项目的设计灵感来自于该地区的工业化背景，这样一方宁静的场所为人们亲密交流提供了方便；不但弥补了邻近社区的功能性缺陷，还改善了街道景观。

公园和周边街道的景观设计同时选用了传统材料与现代材料，例如工业化遗存、观赏植物、蒸馏工艺中所使用的谷物、钢铁和石头等。公园景观包括波浪形的草坪、带有雕饰的混凝土种植槽、传统的碎石墙和现代感十足的琢石墙和水体景观。

该项目对邻近社区而言意义重大，它可以帮助具有前瞻意识的社区居民了解社区周边深厚的历史背景。将景观作为一种艺术媒介，公园的每一处景观都如同被雕刻一般自然：起伏不定的草坪、与众不同的长椅、被狭窄的铁板小径隔开的流线型平滑草地、带镶嵌图案的块石路面……这种极具美感的艺术手段也可在戴维•鲍尔教父车道沿线的街道景观设计中觅得踪影——长椅极具创意地摆设在道路的垂直方向，形成独一无二的标志性景观，整座公园的景致尽收眼底。

合理的树木栽植手法被认为是该项目的一个重头戏。采用结构性土壤来实现树木展幅与生长周期的最优化搭配，这也是该项目设计的关键。在施工前整个场地填埋了底层土，在种植时又追加了五种不同的混合土壤，以确保适宜的园艺生长环境。

Barrel Warehouse Park, a city park in the new downtown core of Uptown Waterloo, is located near historic Seagram distilleries and sits adjacent to two former barrel warehouses that have recently been converted into condominiums. The park's design draws inspiration from the site's industrial qualities and provides opportunities for spontaneous interaction as well as intimate gatherings. It serves to complement the nearby residences, regenerate the landscape, and improve the streetscapes thereby changing the area's ordinarily unwelcoming atmosphere.

A palette of traditional and contemporary materials was used in the design of the park and surrounding streetscape such as several salvaged industrial artifacts, ornamental grasses reminiscent of grains used in the distilling process, steel, and stone. The park consists of undulating lawns, concrete planters with sculpted hedges, traditional rubble-stone walls, and a contemporary architectural cut-stone wall and water feature.

Barrel Warehouse Park is an important contribution to the community because it has helped to create visual connections between the site's current forward-thinking residents and the deeply historical identity of the neighbourhood. This is achieved by using the landscape as an artistic medium where each of the park's elements is very sculptural in nature: rolling lawns, unique benches, flowing grasses bisected by elevated steel catwalks, and mosaic-like stone paving. This artistic approach was also used in the design of the park's streetscape along Father David Bauer Drive, which is uniquely defined by innovative benches that sit unconventionally perpendicular to the road, offering views of the

park rather than the street.

Proper tree planting techniques were a crucial consideration in the completion of this project. Structural soils were used to maximize the size and longevity of the trees, which play a key role in the design. The entire site was also sub-soiled prior to construction and five different soil mixes were used during planting to ensure proper horticultural conditions.

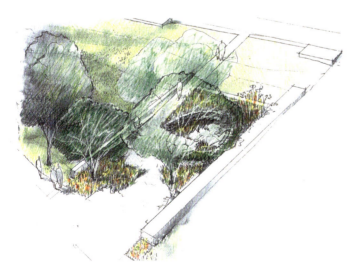

项目位置 : 加拿大安大略省沃特卢市

预　　算 : 65 万美圆

项目时间 : 2000 年～ 2001 年

景观设计 : Janet Rosenberg + Associates （JRA）

客　　户 : 沃特卢市

占地面积 : 5000m²

所获奖项 : 2003 CSLA 地区优秀奖

2004 设计交流银奖

Location: Waterloo, Ontario (Canada)

Budget: $650,000

Project Date: 2000~2001

Landscape Design: Janet Rosenberg + Associates (JRA)

Client: City of Waterloo

Site Size: 5000m²

Awards: 2003 CSLA Award, Regional Merit

2004 Design Exchange Awards, Silver

动态的线性公园——斯科茨代尔滨水区

Dynamic Linear Park—Scottsdale Waterfront

翻译　李沐菲

景观总设计图

仙人球花广场

仙人球花广场入口

仙人球花入口庭院

　　将一期和二期建筑物之间的广场及人行步道（包括亚利桑那运河北岸的建设）整合为一个统一而大胆的主题是该项目的设计主旨。其设计灵感主要来源于沙漠景观以及斯科茨代尔古城的历史风貌。在该设计中，当地仙人掌和仙人球的鲜明轮廓、色彩以及纹理均被体现在广场及各公共空间的多处细节中。

　　该场地部分区域的建筑十分密集，因此设计中添加了高出地面的种植器皿以及矮墙坐椅，以增加土壤深度。阶梯式圆形广场以及中心聚会区都铺设了波状的草坪，为人们提供了更多的休息及野餐区，也成为从运河河畔到建筑群的过渡区域。所有元素共同组成了一个仙人掌的轮廓，而每块草坪则是仙人掌各个多刺的分支。

　　该项目的零售区与运河正面相连，位于项目中心的两栋住宅楼之间。这一空间作为公众的入口区域，以鲜明的步行街为主要特色贯穿于整个项目。北起居住区的下沉广场，其设计仿似一个巨大的树形仙人掌的花朵，而通向运河的小路则如同树形仙人掌的树干。两座著名的艺术品——"门"和"Hashknife 驿马快信"矗立在运河的正面，既表达了对斯科茨代尔历史遗产的敬意也表达了对未来的憧憬。

从休闲娱乐的角度来看，凤凰城中心地区错综复杂的运河网络均未得到充分的利用。从整修一条土层松动的便道开始，该项目将亚利桑那运河流域变成了一座生机盎然的线性公园，包括绿树成荫的公共广场、小型的阶梯式圆形广场、悬臂式运河观景台、幽静的休息区以及两座公共艺术品。该项目是将斯科茨代尔转变成活力四射的城市中心最重要的组成部分。

作为一个四通八达的空间，亚利桑那运河的正面与临近广场的功能区可以作为斯科茨代尔其他地区的公共纽带，而当地居民和出城的游客都把这里当做一个真正的"驿站"。由于该项目属于高密度，许多建筑物下面都建有地下停车场，这为设计团队带来了众多挑战：高密度建筑区与运河河岸该如何过渡，如何保持建筑材料的连贯性以及整个区域的统一性，而平衡及协调各方需求与冲突则是该项目设计的主要成就之一。

An overall concept for Scottsdale Waterfront was developed that transformed the multiple pedestrian plazas and walkways between the buildings of Phase One and Phase Two, including the North Bank of the Arizona Canal into a bold, unifying theme. For inspiration, the concept draws heavily from desert landscape imagery, as well as historical references of Old Town Scottsdale. The concept incorporates the bold forms, colors and textures of native cacti and cactus flowers throughout the plaza and public spaces.

Raised planters and seat-walls were integrated into the design to provide additional soil depth for the trees in the over-structure areas. Curved panels of turf were judiciously used at the amphitheater and central gathering area to provide additional sitting and picnicking areas as a transition between the canal and over-structure areas of the site. When viewed from above, these design features become part of the overall themed cactus forms with turf panels representing pads of the prickly pear cactus.

Retail areas of the project are connected to the canal frontage through the center of the project between the two residential towers. This space was designed as public access to encourage a strong pedestrian linkage through the project. This linkage is anchored on the north by the residential drop-off plaza that is designed to resemble an immense saguaro cactus blossom with the path to the canal forming the saguaro's trunk. Two prominent public art installations bookend the canal frontage, "The Doors" and the "Hashknife Pony Express," pay homage to Scottsdale's heritage and its future.

For years the intricate network of historic working canals throughout the Phoenix Metropolitan area has been under-utilized from a recreational standpoint. Starting with little more than a scarified dirt service road, Scottsdale Waterfront has transformed this stretch of the Arizona Canal into a vibrant linear park including shaded public plazas, a small amphitheater, cantilevered canal

overlooks, quiet sitting areas and two major public art installations. The Arizona Canal at Scottsdale Waterfront is at the heart of the transformation of downtown Scottsdale into a vibrant urban center. As an activated space, the AZ Canal frontage and adjoining plazas function as public linkages to the rest of downtown Scottsdale, designed as a true "destination" for local residents and out-of-town visitors alike. Because of the high density of the project, most of the site and buildings are built on top of the underground parking structures, which created a variety of challenges for the design team to seamlessly transition between the over-structure areas and the canal bank, while maintaining continuity of materials and a strong overall sense of place. Balancing and managing the needs and conflicts presented by this urban infill development was one of the major accomplishments of the site design.

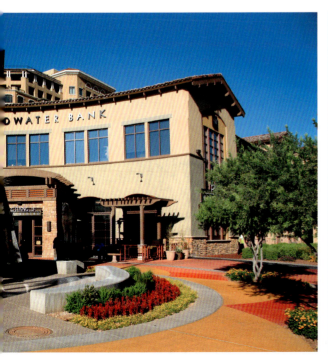

项目位置：美国亚利桑那州斯科茨代尔市

景观设计：Floor 公司

建筑设计：SCB Design｜OPUS 建筑事务所

占地面积：32 374m²（一期 办公区与零售区）
　　　　　22 258m²（二期 SWF 住宅区）
　　　　　1207m（亚利桑那河道改造）

成　　本：75 万美圆（一期）；75 万美圆（二期）；200 万美圆（河道改造）

项目时间：2004 年～ 2005 年（一期）；2005 年～ 2007 年（二期）；2005 年～ 2007 年（亚利桑那河道改造）

所获奖项：2008 年获得了由 Valley Forward 协会颁发的环境设计优秀奖
　　　　　2008 年获得了 AZ|RE RED 超过 10 万个标准单元的最佳多户住宅项目奖

Location: Scottsdale, Arizona, USA

Landscape Architecture: Floor Associates

Architecture: SCB Design|OPUS Architects

Surface Area: 8 Acres (phase 1 offices/retail)
5.5 Acres (phase 2 SWF Residences)
0.75 miles (Arizona Canal Improvements)

Budget: $750,000 (phase 1); $750,000(phase2);
2 million (Canal improvements)

Project Dates: 2004~2005 (phase 1); 2005~2007
(phase 2); 2005~2007 (Canal improvements)

Awards: 2008 Environmental Design Excellence Award Valley
Forward Association Scottsdale Waterfront Residences
2008 Best Multi-Family Project Over 100,000
SF(submitted by Opus West) AZ|RE RED Award

创意无限——奇加科西滑板公园

Great Creativity——Chingacousy Skatepark

翻译　刘建明

该项目堪称具有无限创意的滑板运动与混凝土建材丰富造型之间的一种完美结合。尽管本质上仍是一处滑板广场，但却通过创新理念而推陈出新、独具特色——无论是公园的滑板地形、前卫的欧式建筑，还是设计师"寓于此地"的隐喻式图书馆，无不体现出创新的与众不同。

该项目位于宾顿市的奇加科西公园内，占地面积约1672平方米。独特的造型和细部设计使得项目超凡脱俗——即使场地上没有滑板一族的身影，整个空间也效果非凡。

秉承着对生态负责的态度，该项目完全采用 EcoSmart 生态混凝土建造。这种混凝土含有很高比例的粉煤灰，该项目重新利用了粉煤灰这种废料，但避免其进入填充区中。EcoSmart 生态混凝土也将混凝土生产过程中所固有的二氧化碳排放量降至最低。

虽然混凝土有时被视为一种凝重粗狂的建材，但该项目中运用悬臂、拱形结构以及平拱来展现设计的轻盈、飘逸和富于动感。公园中的一些设计还呼应了埃罗·沙里宁设计的肯尼迪机场和贝特洛·莱伯金设计的伦敦动物园的企鹅池。整座公园充满动感、青春与活力，体现出滑板文化创意无限与蓬勃的朝气。

不同建材打造的独特造型和细部设计贯穿整座公园。独具匠心的木质遮阳篷不仅为炎热的天气带来丝丝凉意，更是年轻滑板族社交聚会的场所。钢质支柱逐渐倾斜变密，彰显出动感十足的韵味。

缎带式突出设计：深受建筑师、艺术家托马斯·赫斯维克（赫斯维克工作室）的钢质缎带式楼梯的启发，设计师将这一设计融入到该项目中，设计了3个连续的、逐渐起伏的隆起，这些隆起又在同一角度的平面上被拉平。它们飘逸起伏，在张弛有度之间展现出动感与力量的释放。

倾斜的"桌面"：悬臂式主题又一次被运用到设计中，

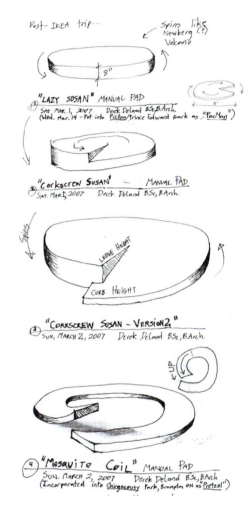

只是这次为了训练平衡技巧而运用到一个圆形平面上。当滑板爱好者踏上圆形"桌面"，悬臂式设计便有效增强了运动的平衡感。一些自行车爱好者选择合适的方式，只是将"桌面"作为他们冲向夜空的一个"发射点"，看到这些高难度的动作，设计师不禁开玩笑似的问外星人在哪里。

漂浮的突出设计：传统滑板公园的滑道都是沉重的灰色石块，它们很容易打断公园景观的流畅性。然而在该项目中，突出部分利用涂了颜色的混凝土作为平拱进行细部处理，这种设计使得此部分独具匠心，而并没有单纯地追求美学效果。公园中玩轮滑、滑板和骑车人此起彼伏、一切都一目了然，

他们的身姿似乎比以前更加轻盈、敏捷。

蜿蜒盘旋式设计：Spectrum 公司的德里克·迪兰德（Derek DeLand）最初的设计草图是一盘蚊香的形式，但是随着设计的进展，吉姆·巴纳姆（Jim Barnum）进一步完善设计并提出了两个彼此偏离卷曲，但又重新会聚成一上一下、设有出口的环形设计。竣工后，这个盘旋式设计将兼具现代雕塑、风火轮赛道和休闲小吃空间的功能。此起彼伏的环形设计形成一系列的明拱，将设计的轻灵与流畅展现得淋漓尽致。

该项目堪称景观设计师、滑板公园设计师、工程师、混凝土生产商和施工方之间通力合作的典范。

The progressive design of Chingacousy Skatepark is a beautiful fusion of the limitless creative possibilities of skateboarding with the infinite formal possibilities of concrete. Essentially a skate plaza, the project achieves distinction by tapping into innovative inspiration, from local skate spots to avant-garde European architecture, to the designers' metaphorical library of "out there" concepts.

Sited in Brampton's Chingacousy Park, this project with approximately 18,000 square feet includes a series of unique features and details that elevate it above the ordinary – the space is so photogenic that it photographs well even without skateboarders on site.

In the interest of ecological responsibility, the park is made entirely of EcoSmart concrete, which contains a high proportion of fly ash, thus repurposing this waste material and keeping it out of the landfills. EcoSmart concrete also helps to minimize CO_2 emissions inherent in concrete production.

Concrete is sometimes seen as a heavy or earthy material, yet here the introduction of themes of air and light via a series of cantilevers, arches, and flat arches instead expresses lightness, skimming, and motion. There are even echoes of Eero Saarinen's JFK Airport or Berthold Lubetkin's Penguin Pool at London Zoo. The overall effect is one of dynamism, youth, and active life, all highly appropriate for the creative energy of skateboard culture.

Special features and detailing appear throughout the park, created in a variety of materials. There is a prominent wood sunshade, casting shadow on hot days and providing a spatial refuge and sense of social gathering for the young participants. In steel, the uprights on the flat bars tilt and accelerate in series, expressing the motion of their use.

The Ribbon Ledges: Inspired by a photo of architect and artist Thomas Heatherwick's (Heatherwick Studio) steel ribbon stairs, the designers ran the idea through skateboarders' minds and came up with three escalating ledges that transition up in series and flatten into the same angled plane. The strong visual momentum of the three-ledge sequence arcs up and out, the energy expressed and released through the cantilevering of the flat bank.

The Tilting Table: The cantilever theme is again expressed, this time incorporated into a round "table" intended for balance tricks. When a skater ollies onto the table into a balance trick, the cantilever enhances the balancing magic of the motion. Some riders take the high road and simply use the cantilever as a blastoff point into the evening sky, inspiring the designing team to playfully ask where E.T. is.

The Floating Ledges: Conventional skatepark ledges have a tendency to be heavy grey blocks interrupting the motion lines of the park landscape. At Chingacousy, ledges are detailed in coloured concrete, as flat arches. Instead of being aesthetic

liabilities, this technique turns the ledges into featured assets. The skatepark's animated movement of legs, boards, and wheels continues without visual interruption; a grind or slide along the ledges appears lighter and quicker than ever.

The Pretzel Rollers: Originally sketched by Spectrum's Derek DeLand as a mosquito-coil form, this element was developed by Jim Barnum through the design process into two diverging, rolling loops that re-converge as an over/under gap. As built, the element is part contemporary sculpture, part Hot Wheels track, and part salty snack treat. The repeating arcs of the rollers have been detailed as an airy series of open arches that express the unbearable lightness of rolling.

Chingacousy Skatepark is a shining example of what can happen when landscape architects, skatepark designers, engineers, concrete contractors and skatepark construction specialists harmoniously integrate their best ideas, materials and workmanship into a single project.

项目位置：加拿大安大略省宾顿市
客　　户：宾顿市政府
景观设计：Spectrum Skatepark Creations 和 LANDinc
建筑摄影：汤姆·阿班
项目预算：550 000 美圆
项目时间：2007 年 8 月 ~ 2008 年 1 月
所获奖项：2008 年安大略省预拌混凝土协会评选的最佳硬质
　　　　　景观设计奖宾顿市城市设计奖

Location: Brampton, Ontario, Canada

Client: City of Brampton

Landscape Design: Spectrum Skatepark Creations with LANDinc.

Architectural Photographer: Tom Arban

Budget: $550,000

Project Dates: August 2007~January 2008

Awards: Ontario Ready Mix Concrete Association 2008 – Best Architectural Hardscape City of Brampton Urban Design Awards

哈得逊河公园

Hudson River Park

翻译　李沐菲

2005 年，哈德逊河公园信托委员会邀请马修·尼尔森 (Mathews Nielsen) 主笔哈德逊河公园的翠贝卡河段部分（第 3 段）的开发设计。这个耗资 1.2 亿美圆的河段项目包括约 56 656 平方米的公共空间和新增建筑物。翠贝卡河段的社区公共空间既可以为人们提供各种娱乐休闲活动的场所，亦可作为公园内滨水步道的延伸区。该项目在当前的废弃码头位置开辟了新的山地公园和两处新的码头延伸结构。

将占地约 11 241 平方米的 26 号码头设计成为河口，能够接触和了解河流生态。码头上有饭店和停船棚屋等建筑，从屋顶露台和泊船区可将广阔的海景一览无遗。

25 号码头一直是码头娱乐组团的核心区域。全新规划的占地面积约 12 821 平方米的码头恢复了各种备受欢迎的娱乐设施，包括沙滩排球场地、迷你高尔夫、儿童游乐场等，这些设施大都采用经久耐用的材料，并保持了相当高的建筑质量水平和艺术性，使得该码头依然是社区居民的挚爱。该码头包括最西边的一个阶梯式的露台观景区域，一个小吃店和码头管理办公室。码头南侧的停泊区可同时停靠 90 艘船。

在 25 号码头和 26 号码头之间，设计有一个占地面积约为 1115 平方米的缓冲平台，这是一处具有多种用途的人型集会广场，主要用于举行音乐会和观看露天电影。该广场也是从翠贝卡邻近社区进入公园的主入口。

26 号码头北侧的高地被设计成波浪形，在木栈道的两侧种植本地植物。此外 32 号码头还是野生动物适宜的栖息地。

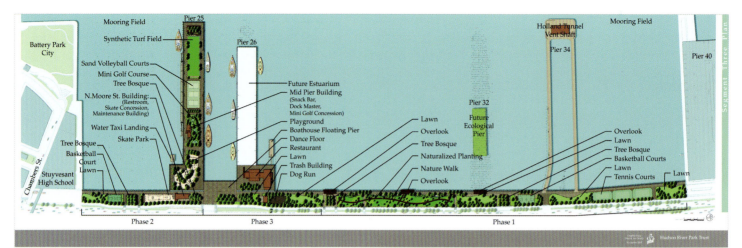

Mooring Field
Pier 25
Synthetic Turf Field
Sand Volleyball Courts
Mini Golf Course
Tree Bosque
N.Moore St. Building: (Restroom, Skate Concession, Maintenance Building)
Water Taxi Landing
Skate Park
Tree Bosque
Basketball Court
Lawn
Stuyvesant High School
Chambers St.
Battery Park City

Pier 26
Future Estuarium
Mid Pier Building (Snack Bar, Dock Master, Mini Golf Concession)
Playground
Boathouse Floating Pier
Dance Floor
Restaurant
Lawn
Trash Building
Dog Run

Lawn
Overlook
Tree Bosque
Naturalized Planting
Nature Walk
Overlook

Pier 32
Future Ecological Pier

Holland Tunnel Vent Shaft
Pier 34
Mooring Field
Pier 40

Overlook
Lawn
Tree Bosque
Basketball Courts
Lawn
Tennis Courts
Lawn

Phase 2   Phase 3   Phase 1

Hudson River Park Trust

In 2005 Mathews Nielsen was engaged by the Hudson River Park Trust to carry forward the development of Tribeca neighborhood portion (Segment 3) of Hudson River Park. This $120 million segment includes 14 acres of open space and new buildings. This significant community open space in Tribeca offers active and passive recreational opportunities while continuing the park-wide signature of the continuous waterfront promenade. The design provides new upland park areas and two new expanded pier structures in the location of the existing derelict piers.

The 121,000 square foot (2.78 acre) pier 26 will house a future estuary offering hands-on learning and engagement with the river ecology. The pier will be home to a restaurant and boathouse building offering dramatic water views from a rooftop deck, and a community boating facility.

Pier 25 has long been an ad hoc agglomeration of recreational activities. The new 138,000 square foot (3.17 acre) pier design reinstates these well-loved venues for sand volleyball, mini-golf and children's play with more durable materials that retain the quality of surprise and artistry that has so endeared the pier to the community. The pier includes an elevated deck viewing area at the far west end and a snack bar building with a dock master's office for a new 90 boat capacity mooring field immediately south of the pier.

Between the piers 25 and 26, a 12,000 square foot relieving platform has been designed with a multi-use plaza for large gatherings such as concerts and outdoor movies. This plaza will serve as the main entrance to the Tribeca neighborhood portion of the park.

The upland immediately north of pier 26 is designed as an undulating topography planted with native grasses experienced from elevated boardwalk. Additionally pier 32 will be planted as a wildlife sanctuary.

项目位置：美国纽约

客　　户：纽约州纽约市哈德逊河公园信托委员会

预　　算：1.2 亿美圆

项目长度：2.4km

项目时间：2005 年 ~ 2008 年

景观设计：纽约马修·尼尔森景观设计事务所

海洋工程：纽约 Halcrow

照明设计：纽约 Fisher Marantz Stone

灌溉设计：北哈芬市 Northern Designs

景观保护：纽约 Li Saltzman Architects

所获奖项：美国景观设计师协会纽约州荣誉奖

Location: New York, USA

Client: Hudson River Park Trust, New York, NY

Budget: US$120 million

Project Length: 2.4km

Project Dates: 2005~2008

Landscape Architect: Mathews Nielsen Landscape Architects PC (New York)

Marine Engineers: Halcrow, New York, NY

Lighting Design: Fisher Marantz Stone, New York, NY

Irrigation Design: Northern Designs, LLC, North Haven, CT

Preservation Architects: Li Saltzman Architects, New York, NY

Awards: American Society of Landscape Architects, New York Chapter, Honor Award

西德威尔友谊中学校园景观

Landscape of Sidwell Friends School

翻译 王玲

西德威尔友谊中学希望在华盛顿特区的中学校园里创造一处既能够体现其办学宗旨，又独具特色、备受欢迎的校园景观。校园的景观设计不仅要强调校方在环境管理方面所做的工作，而且还要突出建筑和景观系统的可视性如何在校园设计中融入教育元素。Andropogon 公司与 Kieran Timberlake 建筑事务所通力合作，共同完成了该项目的总体规划。该规划不仅在宽广的城市肌理中清晰地勾勒出校园的景观设计，同时也为个性化学习创造出亲密优雅的环境。

该项目景观的总体规划和场地设计包括新建的运动区、葱郁的本地植被、新增建筑上的绿色屋顶和一个设有人工湿地的中央庭院。人工湿地为了生态和教育目的处理雨水和污水。设计师巧妙地将水资源管理解决方案融入于景观设计中，从而使得建筑与场地紧密地联系在一起。湿地仿佛是一处"时刻工作着的景观"，它不仅利用生物工艺净化水资源，还向学生生动地展示其工作原理。

这种基于湿地的水处理系统不仅是建筑排放污水的净化器，同时也是学校科学课程生动的教学范本。污水首先在一个地下蓄水池中进行一级处理，然后流经一系列错落有致的芦苇地，因此污水是不可能到达地表的。

附着在石质种植槽和植物根系上的微生物对水中的污染物起到了行之有效的过滤作用。经过处理后的污水具有较高的品质，它们又被重新用于建筑的厕所冲水和冷却塔的补充水。相较于传统的水处理系统，这种基于湿地的水处理系统更加节能高效，它的每一个环节都是利用环保的生物工艺，而不是投放化学品。

该项目的另一个特点是将雨水引入到雨水花园和永久生物池塘中。屋顶雨水被收集到一个地下蓄水池，干旱时可将蓄水池中的水用于保持池塘水位；蓄水池多余的雨水则从一个磨石形成的"泉眼"里涌出，再注入池塘中。雨量较多时，

池塘里的水通过一个带槽的堰溢入雨水花园，这也是模仿漫滩环境的一种做法。

植物景观在微观上遵循不同植物群落的特点，按照土壤水分梯度自然分布。绿色屋顶上设置了用于城市农业研究的种植槽和用于艺术课的修剪园。多数硬质景观都由回收再利用材料构成。再生石在该项目中被广泛使用，人行道采用再生石板，木栈道则是从巴尔的摩港回收的绿心樟木桩制成。

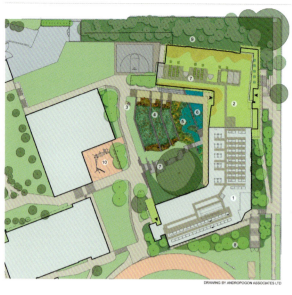

1. EXISTING MIDDLE SCHOOL
2. MIDDLE SCHOOL ADDITION WITH GREEN ROOF
3. TRICKLE FILTER WITH INTERPRETIVE DISPLAY
4. WETLANDS FOR WASTEWATER TREATMENT
5. RAIN GARDEN
6. POND
7. OUTDOOR CLASSROOM
8. BUTTERFLY MEADOW
9. WOODLAND SCREEN AT NEIGHBORHOOD EDGE
10. PLAY EQUIPMENT

DRAWING BY ANDROPOGON ASSOCIATES LTD

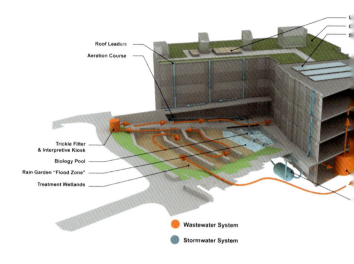

Roof Leaders
Aeration Course
Trickle Filter & Interpretive Kiosk
Biology Pool
Rain Garden "Flood Zone"
Treatment Wetlands

● Wastewater System
● Stormwater System

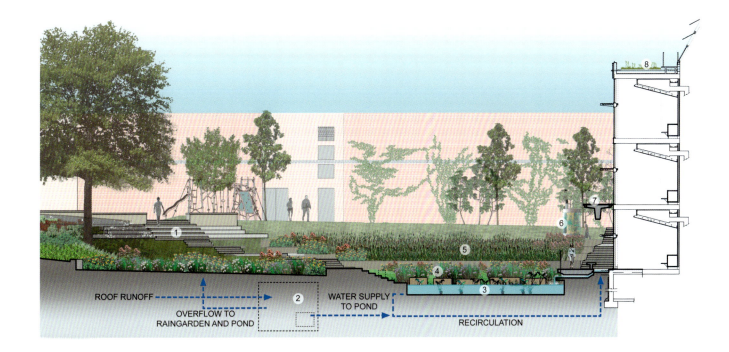

ing Beds

Tank

t Tanks and Filters
Water Storage

r Cistern

ROOF RUNOFF ---------> ② WATER SUPPLY
OVERFLOW TO TO POND
RAINGARDEN AND POND RECIRCULATION

Sidwell Friends School sought to create a campus landscape that maintained Quaker values while presenting a distinctive and welcoming aesthetic for their Middle School Campus in Washington D.C. There is a strong emphasis on demonstrating environmental stewardship, on making building and landscape systems visible, and on the integration of educational opportunities into the campus design. Andropogon collaborated with architects Kieran Timberlake Associates to develop a Master Plan for articulating the campus landscape within the larger urban grid, while creating intimate environments for individualized learning.

Andropogon's Landscape Master Plan & Site Design included new play areas, native plantings to provide screening for neighbors, a green roof on the new building addition, and a central courtyard with a constructed wetland designed to utilize storm and wastewater for both ecological and educational purposes. Andropogon's plan integrated water management solutions into the landscape, inextricably linking the building to its site. The wetland becomes a "working landscape"; using biological processes to clean water while providing students with a vivid example of how such systems work in nature.

The wetland-based treatment system, for the cleaning of wastewater and gray water from the building, is envisioned as a teaching tool for the science curriculum. Wastewater is given primary treatment in an underground tank then circulated through a series of terraced reed beds; wastewater is not accessible at the surface.

Microorganisms attached to the stone planting media and the plant roots provide an efficient breakdown of water contaminants. The high-quality outflow from the system is recycled into the building for reuse in toilet flushing and make up water for the cooling tower. The operational and energy efficiency of this wetlands-based system is very high compared to most conventional treatment systems. Each component relies on biological processes rather than chemical inputs.

An additional feature of the site design directs storm water runoff to a rain garden and permanent biology pond. Runoff from the roof is collected in an underground cistern that maintains the water levels of the pond during dry weather. Excess rainwater in the cistern wells up from a millstone "spring", then flowing to the pond. During a heavy rain, the pond will also overflow through a slotted weir into the rain garden which mimics the performance of a floodplain environment.

The planting design follows, at a micro scale, the range of plant communities that would naturally occur along this soil moisture gradient. The green roof incorporates planters for urban agriculture projects and a cutting garden for art classes. Most of the hard elements in the landscape are reclaimed or recycled. Material for the walkways is recycled flagstone and decks are wood from green heart lumber pilings reclaimed from the Baltimore Harbor. There is also extensive use of reclaimed stone.

项目位置：美国华盛顿特区
成　　本：2850 万美圆
项目时间：2003 年～ 2006 年
景观设计：Andropogon Associates（宾夕法尼亚州费城）
建　筑　师：Kieran Timberlake Associates（宾夕法尼亚州费城）
占地面积：60 702.9m²
建筑面积：6688.8m²
所获奖项：2007 年美国建筑师协会环保委员会十大绿色建筑奖
　　　　　2007 年美国绿色建筑协会 LEED（能源与环境设计先锋）白金认证

Location: Washington, D.C. (USA)

Budget: $28,500,000

Project Dates: 2003~2006

Landscape Design: Andropogon Associates, Ltd. (Philadelphia, Pennsylvania)

Architect: Kieran Timberlake Associates, LLP, (Philadelphia, Pennsylvania)

Site Size: 60,702.9m²

Building Area: 6,688.8m²

Awards: 2007 AIA COTE Top Ten Green Projects Selected by The American

Institute of Architects, Committee on the Environment.

2007 Certified LEED™ Platinum (Leadership in Energy & Environmental

Design) by the U.S. Green Building Council

线性公园——卡罗溪公园

Linear Park—Carroll Creek Park

翻译 王玲

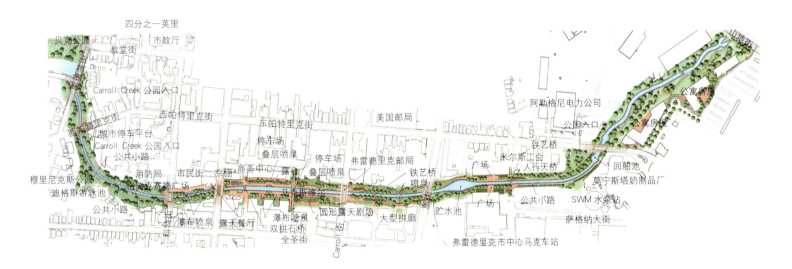

四分之一英里
贝克公园   市政厅
教堂街
Carroll Creek 公园入口
马里兰克街   西帕特里克街   东帕特里克街   美国邮局
城市停车平台
Carroll Creek 公园入口   停车场
公共小路   停车场   弗雷德里克邮局
叠层喷泉
消防局   市民街   凉棚   商务中心   露台   叠层喷泉   铁艺桥
穆里尼克斯公园   防洪赛德广场   喷泉
迪格斯游泳池   Carroll 街   圆形露天剧场   大型拱廊
公共小路   瀑布喷泉   露天餐厅   双拱石桥   全圣街   弗雷德里克市中心马克车站
阿勒格尼电力公司
公园六口
铁艺桥
米尔斯工会   公寓房屋
广场   人行天桥   寓房屋
回船池
贮水池   莫宁斯塔奶制品厂
广场   公共小路   SWM 水泵站
萨格纳大街

该项目位于美国东北部马里兰州费雷德里克市区，是城市振兴的获奖项目，也是 HNTB 建筑设计事务所面临的一个非同寻常的挑战。项目要求在一条宽敞的混凝土水道上建造充满活力的公共空间。这条混凝土水道十几年来一直是马里兰州最大的历史区——费雷德里克市区的一个棘手问题。尽管该项目还未完全竣工，但它已荣获 2007 年美国规划协会马里兰州年度奖和 2008 年国际经理理事会奖。

费雷德里克市经济发展委员会是该项目区域发展提案的积极支持者，该项目也是费雷德里克市区著名的多功能城市公园。费雷德里克市历经多年的经济衰退和洪水泛滥，建设该项目并建立相关防洪设施成为促进雷德里克市的城市振兴和经济发展的一项大胆尝试。作为一个多功能综合区，该项目利用开发为附近区域带来了 1.55 亿美圆的收益。尽管经过了 30 多年的历史沧桑，该项目却仍是利用公园和水道作为地区振兴催化剂的成功典范。

费雷德里克市政府于 2003 年委任 HNTB 建筑设计事务所对该项目进行概念设计、施工图设计以及建设运营和管理工作。HNTB 建筑设计事务所与马里兰州巴尔的摩市的 RK&K 工程公司合作，于 2006 年共同完成该项目的市区一期建设，其中包括与新开发设施和再利用设施同时设计的红砖广场、坡道、坐席和桥梁。

人行道旁的格架覆盖着的咖啡座、两座户外广场、一条商业街和多处喷泉共同展现着该项目的别样风貌。这些喷泉不仅创造出生机勃勃的景观，还有效地补给了河道用水；有的喷泉构成叠层喷泉流入河道，有的只是在路旁或河道水池里的简单喷泉。二座人行桥将河道两岸连接起来并展现出建筑的美感——其中一座是体现当地历史传统的双拱石桥，另一座是由当地艺术家建造的铁艺桥，最后一座人行桥采用了当地建筑中的阶梯状砖立面元素。双向人行系统保障了沿水道的道路，商铺前的道路以及连接它们的道路和水道两岸的桥梁的无障碍通行。该项目设计在风格、建筑规模、历史传承、土地利用类型和强度、新开发设施及其与市区和周边地区的联系方面均有所突破。

设计的核心部分是一座连接市区南北两区的、现代感十足的人行天桥。设计团队在设计过程中面临着许多场地和规划上的挑战——即桥梁跨度大，其可行的支撑点又很少；尽量减小桥的跨度；在北桥墩设置公园入口；避开诸多的公用设施等。于是设计师采用创新手法，通过一个将桥面一分为二的铁架结构支撑桥面，铁架则利用中央支柱垂吊的绳索拉力而固定。竣工后的人行天桥美观实用，成为费雷德里克市的一处新地标。

卡罗溪公园对促进费雷德里克市区的振兴起到了非常重要的作用。截至目前，对公园的再投资已超过 3 亿美圆。其中 1.5 亿美圆投资的一部分正在筹划中，另一部分已经投入建设。该项目一期建设的成功使得费雷德里克市 2007 年获得交通改善基金会 300 万美圆的资金，这也是马里兰州历史上获得的最大一笔基金。该项目二期建设增加了 2500 米的自行车道、一座桥、四个喷泉和由格架覆盖的户外就餐区。

Carroll Creek Park is an award-winning urban revitalization project in historic downtown Frederick, Maryland. The project presented an unusual design challenge for HNTB – to create a vibrant public amenity on top of a large concrete conduit that was a decades-long eyesore in downtown Frederick, the largest historic district in the state of Maryland, in the northeastern United States. The project, which is only partially complete, was recognized by the Maryland APA chapter as its 2007 Project of the Year and by a 2008 International Economic Development Council award (IEDC).

The City of Frederick Department of Economic Development was a clear standout with the Carroll Creek Park Neighborhood Development Initiative. Carroll Creek Park is a world-class mixed-use urban park through historic downtown Frederick. The park and related flood control project was a bold initiative of the City of Frederick to encourage downtown revitalization and economic development following years of decline and devastating floods. A fully mixed-use neighborhood, Carroll Creek Park is generating US$155 million in adjacent infill and adaptive reuse development. This project, which now spans 30 years and five administrations, is an excellent example of using a park and waterway as a catalyst for neighborhood revitalization.

The City of Frederick selected HNTB in 2003 for concept design of the park, construction documentation, and construction administration and management. Teamed with RK&K Engineers in Baltimore, Maryland, the HNTB team completed the initial downtown phase in 2006, consisting of brick plazas, ramps, seating areas and bridges designed in tandem with new development and adaptive reuse structures. The HNTB team is currently designing the remaining phase of the park and will complete bid documentation in early 2009, with second phase implementation scheduled to complete in 2010.

The park features sidewalk cafes under trellises, two outdoor amphitheaters, a market pergola, and several fountains. Some fountains create waterfall cascades into the creek, while others are simply jets of water emerging from the pavement or pools

within the creek. The fountains create excitement and have the added benefit of aerating the creek water. Three pedestrian bridges provide architectural interest and access across the creek. Each bridge has a unique design focus: one is a double arched stone bridge based on historic bridges in the region, a second is an ornamental iron bridge constructed by a local artist and the third utilizes a stepped brick façade reminiscent of the local architecture. A two tier pedestrian system was included to provide ADA accessibility along the creek, to businesses facing the creek, between the two paths, and across the creek. The design responds to changes in style and scale of architecture, historic context, type and intensity of land uses, new development, and linkages to downtown and adjacent neighborhoods.

A centerpiece component of the design is a state-of-the-art suspension pedestrian bridge linking the north and south side downtown neighborhoods. HNTB's New York bridge architecture office faced a variety of site and programmatic challenges for the design, including a long span length with few feasible locations for supports and the need to minimize deck depth, accommodation of a pedestrian entrance to the park at the north abutment, and the necessity of avoiding numerous utilities. The HNTB designers responded with a unique, innovative approach of a bifurcated deck suspended from a center mast support. The final product is a highly functional design that also provides a new landmark for the City of Frederick.

Carroll Creek Park has been an extremely important catalyst for revitalization of Frederick's downtown with an assessed reinvestment value of over US$300 million to date. More than US$150 million in investments are currently planned or under construction. The success of the initial phase of construction resulted in the City being awarded a Transportation Enhancement Grant in 2007 for an amount of US$3 million, one of the largest in Maryland's history. Phase 2 will complete a further 2.5 km (1.5 miles) of bike trails, add another bridge, four new fountains and an outdoor trellised dining space.

项目位置：美国马里兰州费雷德里克
客　　户：费雷德里克市
项目预算：约 5 亿美圆（一期）
长　　度：2 千米（一期）；2.5 千米（二期）
施工时间：2006 年（一期）；2009 年～ 2010 年（二期）
建筑／景观设计：HNTB 建筑设计事务所
所获奖项：2007 年美国规划协会马里兰州年度奖
　　　　　2008 年设计优秀奖（国家公司）
　　　　　2008 年国际经济理事会奖

Location: Frederick, Maryland, USA

Client: City of Frederick

Budget: approx. US$5 mil. (Phase 1)

Length: 2 km (Phase1); 2.5 km (Phase 2)

Project Dates: 2006 (Phase 1); 2009~2010 (Phase 2)

Architects/Landscape Architect: HNTB Architects

Awards: 2007 Project of the Year, Maryland APA

　　　Design Excellence Award 2008 (by the national corporation)

　　　2008 International Economic Council Award

长岛绿城——银杯工作室

Long Island City—Silvercup Studio

翻译　李沐菲

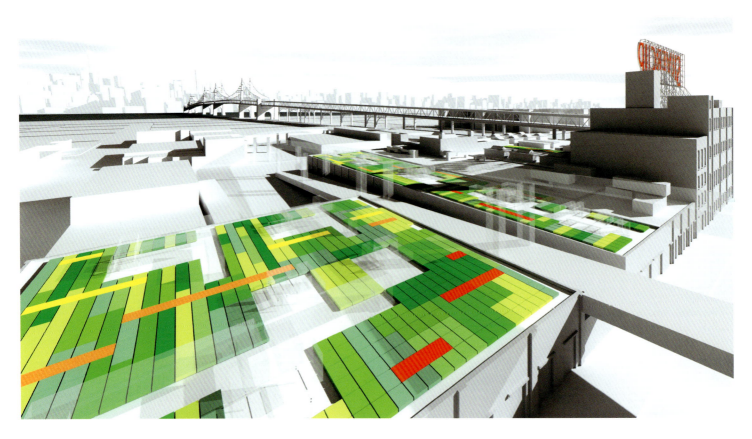

　　绿色屋顶不仅为城市景观的革新创造了独特的未来，还缓解了危害城市的环境问题。戴安娜·巴尔莫里将快速扩展的城市屋顶称为"第五立面"，对于设计师和建筑师来说，这都是一个崭新的领域。城市屋顶不仅可以改变建筑的外观，还可以改变建筑的功能。

　　该项目是巴尔莫里建筑事务所（Balmori Associates）规划的"长岛绿城"中的第一个示范项目，"长岛绿城"规划的主要内容便是在这一工业与住宅综合区内创建一个绿色屋顶网络，其总覆盖面积约 3252 平方米，是目前纽约市最大的绿色屋顶，也是第一个进行科学监控的绿色屋顶项目。

　　该项目将会抑制空气中污染物的扩散并释放氧气，回收超出城市排水设施负荷的降水，还能起到隔热的作用，并以此减少建筑供暖及制冷所需的能源消耗。

　　设计采用的植被组合由 20 多种景天属植物构成，多层次的绿色植物形成丰富的线条和巨大的几何图案。从春季到秋季，这些图案也随着植物的花期变幻着色彩，包括黄色、红色和粉色。

　　该项目通过巴尔莫里建筑事务所（Balmori Associates）、银杯工作室、长岛市商业发展部（LICBDC）以及空气净化联合会的公私合作得以实现；"空气净化联合会"是一个由东

北部各州空气净化中心以及空气使用协调管理协会共同发起的组织（NESCCAF/NESCAUM），旨在为纽约低收入社区实施减少空气污染及节约能源的计划。该项目的设计研究还得到了地球宣言组织（Earth Pledge）、联合爱迪生电力公司以及布拉特学院社区及环境中心（PICCED）的大力支持。

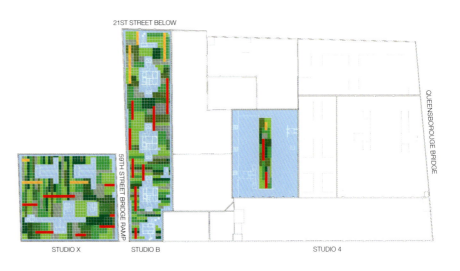

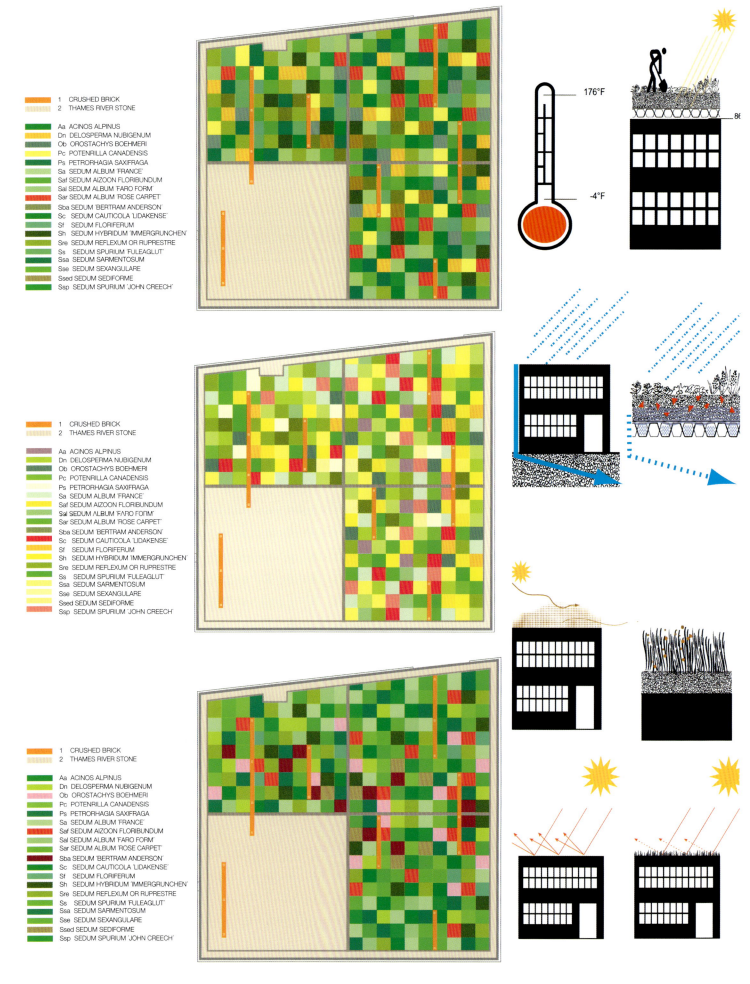

1 CRUSHED BRICK
2 THAMES RIVER STONE

Aa ACINOS ALPINUS
Dn DELOSPERMA NUBIGENUM
Ob OROSTACHYS BOEHMERI
Pc POTENRILLA CANADENSIS
Ps PETRORHAGIA SAXIFRAGA
Sa SEDUM ALBUM 'FRANCE'
Saf SEDUM AIZOON FLORIBUNDUM
Sal SEDUM ALBUM 'FARO FORM'
Sar SEDUM ALBUM 'ROSE CARPET'
Sba SEDUM 'BERTRAM ANDERSON'
Sc SEDUM CAUTICOLA 'LIDAKENSE'
Sf SEDUM FLORIFERUM
Sh SEDUM HYBRIDUM 'IMMERGRUNCHEN'
Sre SEDUM REFLEXUM OR RUPRESTRE
Ss SEDUM SPURIUM 'FULEAGLUT'
Ssa SEDUM SARMENTOSUM
Sse SEDUM SEXANGULARE
Ssed SEDUM SEDIFORME
Ssp SEDUM SPURIUM 'JOHN CREECH'

176°F

-4°F

1 CRUSHED BRICK
2 THAMES RIVER STONE

Aa ACINOS ALPINUS
Dn DELOSPERMA NUBIGENUM
Ob OROSTACHYS BOEHMERI
Pc POTENRILLA CANADENSIS
Ps PETRORHAGIA SAXIFRAGA
Sa SEDUM ALBUM 'FRANCE'
Saf SEDUM AIZOON FLORIBUNDUM
Sal SEDUM ALBUM 'FARO FORM'
Sar SEDUM ALBUM 'ROSE CARPET'
Sba SEDUM 'BERTRAM ANDERSON'
Sc SEDUM CAUTICOLA 'LIDAKENSE'
Sf SEDUM FLORIFERUM
Sh SEDUM HYBRIDUM 'IMMERGRUNCHEN'
Sre SEDUM REFLEXUM OR RUPRESTRE
Ss SEDUM SPURIUM 'FULEAGLUT'
Ssa SEDUM SARMENTOSUM
Sse SEDUM SEXANGULARE
Ssed SEDUM SEDIFORME
Ssp SEDUM SPURIUM 'JOHN CREECH'

1 CRUSHED BRICK
2 THAMES RIVER STONE

Aa ACINOS ALPINUS
Dn DELOSPERMA NUBIGENUM
Ob OROSTACHYS BOEHMERI
Pc POTENRILLA CANADENSIS
Ps PETRORHAGIA SAXIFRAGA
Sa SEDUM ALBUM 'FRANCE'
Saf SEDUM AIZOON FLORIBUNDUM
Sal SEDUM ALBUM 'FARO FORM'
Sar SEDUM ALBUM 'ROSE CARPET'
Sba SEDUM 'BERTRAM ANDERSON'
Sc SEDUM CAUTICOLA 'LIDAKENSE'
Sf SEDUM FLORIFERUM
Sh SEDUM HYBRIDUM 'IMMERGRUNCHEN'
Sre SEDUM REFLEXUM OR RUPRESTRE
Ss SEDUM SPURIUM 'FULEAGLUT'
Ssa SEDUM SARMENTOSUM
Sse SEDUM SEXANGULARE
Ssed SEDUM SEDIFORME
Ssp SEDUM SPURIUM 'JOHN CREECH'

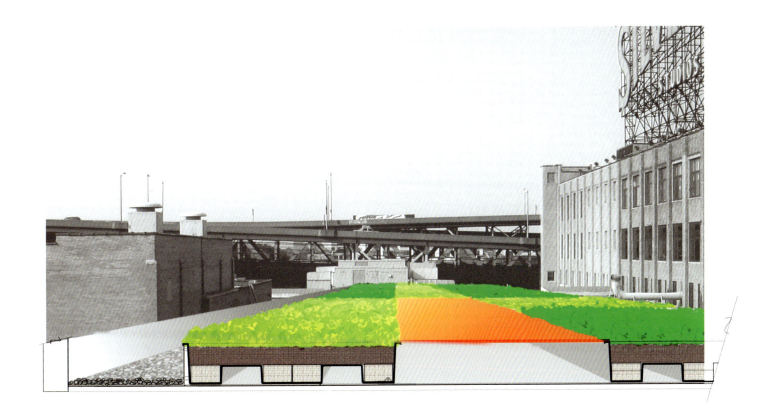

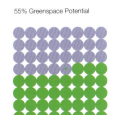

55% Greenspace Potential

Totak Possible Green Area 26,678,743ft^2
(55%) 667 acres

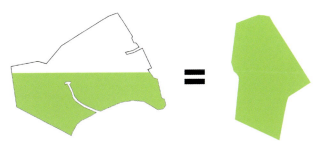

Green Potential Area in Long Island City = Area of Prospect Park

Green roofs offer an alternative future in the evolution of the urban landscape with the ability to alleviate some of the environmental problems that plague our cities. This "fifth facade", as Diana Balmori has called the huge expanses of urban rooftops, is a new frontier for designers and architects where we can change not only the way a building looks, but also how it functions.

The Silvercup Studios green roof is the first demonstration project in Balmori Associates "Long Island (Green) City" proposal to create a network of green roofs in this mixed industrial and residential neighborhood. Covering just over 35,000 square feet of rooftop space, it is the largest green roof ever installed in New York City and the first to be scientifically monitored.

The Silvercup green roof will trap pollutants in the air and release oxygen, absorb storm water that as run-off overburdens the City's sewer infrastructure, and insulate the building reducing its heating and cooling needs and energy costs. Balmori Associates selected a plant palette of 20 different sedum varieties and organized the different hues of green and textures of leaves into linear strips and larger geometric blocks. The pattern will constantly be changing as the sedums bloom in yellows, reds, and pinks at different times from the spring through the fall.

This project was made possible through a public/private partnership of Balmori Associates, Silvercup Studios, the Long Island City Business Development Corporation (LICBDC) and Clean Air Communities (a program, initiated by Northeast States Center for a Clean Air Future and Northeast States for Coordinated Air Use Management (NESCCAF/NESCAUM), committed to implementing air pollution reduction and energy efficiency strategies in low-income New York City communities). Additional support came from the Green Roof Infrastructure Study with Earth Pledge, Con Edison, and the Pratt Institute Center for Community and Environmental Development (PICCED).

项目位置：纽约长岛市
占地面积：约 3252 平方米
建成时间：2005 年
客　　户：银杯工作室
设计团队：巴尔莫里建筑事务所
　　　　　莎拉特建筑事务所

Location: Long Island City, New York
Site Size: approx. 3,252m²
Completed Date: 2005
Client: Silvercup Studios, Inc.
Design Team: Balmori Associates, Inc.
　　　　　　 Shalat Architects P.C.

多功能户外空间——梅萨艺术中心

Multifunctional Outdoor Space—Mesa Arts Center

翻译　王玲

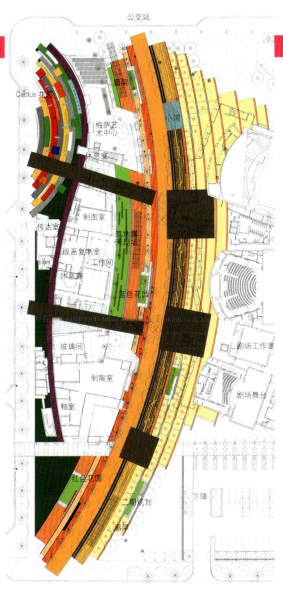

　　尽管梅萨市发展迅速，但却没有城市地标性建筑。政府领导也意识到了这个问题，为了保持目前的人口规模，吸引未来潜在的定居者，梅萨市必须通过有效的城市重建来塑造新形象。城市改造的核心是建造一座满足社区艺术的设施，包括四座表演艺术剧院、一座当代艺术画廊和一座艺术教育中心。

　　在该项目的设计中，社区负责人希望打造一个标志性的公共户外空间。除了可以在这里举办一些城市范围内的大型活动，也能满足小型私人聚会对空间的需求，同时还希望通过创建城市目的地和沽动中心来促进整个梅萨市区的发展。最终，该场地将成为梅萨市的城市地标，不仅在梅萨市的市民心中，而且在凤凰城的市民心中提升了梅萨市的城市形象。玛莎·施瓦兹合伙人事务所和BOORA建筑设计事务所与社区负责人通力合作，共同打造一处无论在视觉还是社会影响方面都极具吸引力的艺术中心。

　　经过对若干不同设计方案的研究，设计师最终决定采用"晶洞"这一设计概念。一个城市密集区沿着场地的四周建立起来，场地内则是主题为"水晶宝石"的开阔空间，此外，设计还考虑到亚利桑那州炎热干燥的沙漠气候、高强度日照以及人们的遮阳需求等因素。

　　一条粗犷的弧形"林阴步道"贯穿于剧院、艺术学校和画廊之间，引导人们进入场地。沿步道分布的空间错落分布、各具特色，为人们提供了休闲放松、艺术展览、非正式表演、大型聚会或简单聚餐会友的空间。摇曳婆娑的树木沿着步道生长，在地面上形成生动的光影效果。交错重叠的不锈钢遮阳篷和彩色玻璃屏也在地面上洒下彩色光影，精心挑选的本地仙人掌和水景的设计均体现出可持续性设计在美国西南部的需求。

　　与林阴步道平行的是一条"旱谷"，这是美国西南部地区一种巨石填沟的特有水景。由人工切割的石阶和火山岩组成的"旱谷"坐落在林阴步道和剧院之间，河床上有时也会有水断断续续地流过，这不禁使人们联想到当地易骤发洪水的地域特点。河床急流上的小桥充当了通向剧院的道路，人们跨过小桥参加剧院的各种活动。急流过后，河床干枯，等待着下次水流的循环。

　　林阴步道的另一个水的主题元素是"宴会桌"，它在艺术中心的设计中曾多次出现。狭长的不锈钢"宴会桌"内设一个流水槽，同时，"宴会桌"也寓意各地宾朋可以汇聚于此。

　　夜幕降临，梅萨艺术中心熠熠生辉，分外妖娆。"宴会桌"底部被灯光装点，玻璃砖墙也披上了彩色的外衣。半透明的天窗光彩明亮，从里面变换着色彩。水流从灯光装饰的出水管里流入河床，灯光透过剧院的玻璃墙射进剧院中。

　　梅萨艺术中心不仅诗情画意地诠释了环境特色，而且为艺术欣赏创造了空间，树立起了城市中心的新形象。该艺术中心自2005年9月正式对外开放以来，已成为了推动当地旅游业和税收发展的重要引擎。

Martha Schwartz Partners began designing a landscape for Mesa Arts Center in Mesa, Arizona in 1999. As the City of Mesa had been developing rapidly without a center or identity, the city leaders realized that in order to retain a population and attract future inhabitants they needed to reinvent Mesa through substantial urban regeneration. Central to this regeneration, a new facility was planned that would house an expansive community arts program including four performing arts theaters, a contemporary art gallery, and an arts education center.

For Mesa Arts Center, the community project leadership envisioned a signature public outdoor space that would host shared, city-wide events while also offering a public space for intimate personal experiences like picnics and promenades. The design also needed to catalyze the redevelopment of downtown Mesa by providing a destination point and a recognizable sense of vitality to Mesa's core. Finally, the complex needed to achieve the status of community icon, improving the city's identity in the minds of its own citizens and the citizens of the broader Phoenix metro area. The design team of Martha Schwartz Partners and BOORA Architects, Inc. worked collaboratively with community leaders to design a visually and socially appealing site for the arts.

After exploring a number of different schemes, a "geode" concept was developed in which an urban density was established along the site's perimeter while leaving a "crystalline jewel" of open space within the site. Design for the site was further informed by the hot dry climate of the Arizona desert landscape, the desire for shade, and the intense character of the sunlight.

As now constructed, a bold, arcing "Shadow Walk" promenade draws people into the site and moves between the Center's theaters, art school, and gallery. Numerous overlapping

spaces along the promenade, all of differing character, provide opportunities for quiet relaxation, art exhibits, informal performances, large gatherings, or just having lunch. Long curving lines of trees, each with their unique shadow patterns, shift back and forth along the Shadow Walk, creating a cadenced, yet dynamic, interplay of light and shadow. Woven stainless steel canopies and colored glass screens cast colored shadows on the ground. Selective use of native cacti and water features reflects the need for sustainable practices in the American southwest. Paralleling the Shadow Walk is the "Arroyo," a water feature whose narrative is appropriate to the boulder-filled ditch of that name common to the southwest. Here, the Arroyo, made of cut stone steps and lava rock, runs along the entire length of the Shadow Walk and lies between it and the theaters. Water is released intermittently and rushes down the riverbed recalling the flash floods characteristic of the region. Entry walks to the theaters bridge the Arroyo torrent giving dramatic access to events. After the flood, the water empties and the cycle eventually begin again. Another water motif running through the Shadow Walk is that of the Banquet Table, an element that appears several times

throughout the plaza. The Banquet Tables are long, stainless steel tables with a running water slot. They invite people to lean and touch and provide a gathering place near the theaters.

At night, Mesa Arts Center glows invitingly. The Banquet Tables are lit from below. Glass block walls are backlit with color. The translucent skylight boxes glow and change colors from within. Water gushes down the Arroyo from dramatically lit outflow pipes.

Light spills invitingly from the glass walls of the theaters.

While Mesa Arts Center creates a new image in the heart of the City, the plaza design poetically addresses the environment and brings the community together in celebration of the arts. The Center, which has been open to the public since September 2005, is already a destination point within Mesa generating tourism and revenue for the City.

项目位置：美国亚利桑那州梅萨市

占地面积：约 28 328 平方米

客　户：梅萨市政府

预　算：9790 万美圆（整个项目）；750 万美圆（景观设计）

项目时间：1999 年～2005 年

景观设计：玛莎·施瓦兹合伙人事务所

建筑设计：BOORA 建筑设计事务所（俄勒冈州波特兰市）

执行建筑师：DWL Architects + Planners（亚利桑那州凤凰城）

记录景观设计师：Design Workshop（亚利桑那州坦佩市）

技术景观顾问：Ryan Associates（马萨诸塞州沃尔瑟姆市）

所获奖项：2006 年美国城市土地协会优秀奖
　　　　　2007 年美国景观设计协会专业设计荣誉奖

Location: Mesa, Arizona, USA

Site Size: approx. 28,328m²

Client: City of Mesa

Budget: $97,900,000 (total project); $7,500,000 (landscape only)

Project Dates: 1999~2005

Landscape Architect: Martha Schwartz Partners

Architect: BOORA Architects (Portland, Oregon)

Executive Architect: DWL Architects + Planners (Phoenix, Arizona)

Landscape Architect of Record: Design Workshop (Tempe Arizona)

Technical Landscape Consultants: Ryan Associates (Waltham, Massachusetts)

Awards: Urban Land Institute, Award for Excellence, 2006
　　　　American Society of Landscape Architects, General Design Honor Award, 2007

国家港口
National Harbor

翻译　李沐菲

国家港口是沿波托马克河而建的一片综合建筑群，坐落在马里兰州华盛顿特区乔治王子郡以南。考虑到首都华盛顿各主要旅游区的交通状况，国家港口被构思成旅游和召开会议的场所，以替代华盛顿那种完全的都市体验。尽管该地区的开发设计和构造灵感依然来自于包括华盛顿、安纳波利斯和巴尔的摩内港的大区域。佐佐木的角色是外部主要空间的规划师兼景观设计师，也是标识系统的设计者及主要建筑物的建筑设计师。庞大的项目规模、预计将超过20亿美圆的预算以及开发商宏伟的目标，这些都意味着将有众多的公司参与合作来共同实现这一目标。

一条被称为中央大街的主干道由两排悬铃木和中间的主道构成，其灵感来自于巴塞罗那著名的兰布拉斯大道。中央大街的地面铺装颜色和纹理与波托马克河天然沙滩自然衔接，这条大街和一个主要滨水区广场被设计为举行庆典、演出及其他活动的场地。在该项目中，所有交通工具的入口被称做"国家大门"，游客将穿过一片白桦林，然后到达以阿尔伯特·帕雷（Albert Paley）所设计的雕塑作品《心动》为标志的入口。

盖洛德国家旅游及会议中心（该建筑不是由佐佐木设计的）于2008年作为国家港口的中心区域开放，这是东海岸最大的非赌博性酒店及会议中心。在正常人流的情况下，这里的设施可以确保参观游客的需求。在白天办公区域内可满足大量人员流动的需求，而住宅区则给人以完善、安逸的社区街道的感觉。

佐佐木的标识系统设计策略主要通过多彩的亭子、路标和指示牌来帮助游客或该项目周围的新居民识别道路，所有的图形都是为了记录这个地方的历史。以亭子为例，采用蓝色石板作为主体，上端配以不锈钢缆绳连接的玻璃顶，以此来象征当地曾经是波托马克河滨港口的历史。

National Harbor is a mixed-use complex located along the Potomac River just south of Washington, DC in Prince George's County, Maryland. While convenient to key tourist sites in the Nation's Capital Region, National Harbor is conceived as a resort and convention destination that offers an alternative to the strictly urban experience of Washington proper. Nonetheless the urban design and architecture of the development draws inspiration from the great urban places in the region including Georgetown, Annapolis and the Baltimore Inner Harbor. Sasaki's role is as urban planner and landscape architect for the principal exterior spaces, designer of graphics and way-finding and as architect of certain key buildings. The huge scale of the project–estimated at over $2 billion – and the ambition of the developer, the Peterson Companies, have meant numerous firms collaborating to realize the vision.

A major thoroughfare called Grand Avenue is defined by an allée of plane trees and establishes the primary spine and is inspired by the famed Ramblas in Barcelona. The color and texture of the avenue's paving set the stage for its terminus – a natural sandy beach along the Potomac. Both this avenue and a main waterfront plaza are meant to accommodate festivals, performances and other events. On the plaza, numerous retail storefronts promote street activity and urban interaction. At the project's vehicular entrance – called the National Gateway – motorists will pass through a grove of birch trees and arrive at a gateway portal featuring a site-specific sculpture, "The Beckoning", by Albert Paley.

The Gaylord National Resort and Convention Center (building not designed by Sasaki) opened in 2008 as the anchor to National Harbor. It is the largest non-gaming hotel and convention center located along the Eastern Seaboard. This facility will ensure a regular flow of transient visitors to National Harbor, the office component will mean a large daytime commuter population, and the residential portions of the site will give it a sense of an established, urbane neighborhood. Sasaki's Graphics/Wayfinding strategy includes colorful kiosks, street markers and directional signs to help both visitors and new residents around National Harbor. All of the graphics are meant to evoke the history of the site. Kiosks, for example, feature slate blue topped by glass "portals" which stainless steel fitting cables as a reference to the site's history as a riverfront harbor of the Potomac.

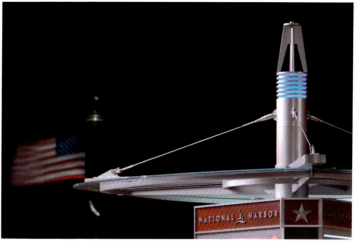

项目位置：美国马里兰州乔治王子郡
占地面积：约 1 214 062 平方米
客　　户：Peterson Companies
预　　算：4300 万美圆
景观设计：佐佐木建筑师事务所
建成时间：2008 年

Location: Prince George's County, Maryland, USA
Site Size: approx. 1,214,062m²
Client: Peterson Companies
Budget: US$43 mil.
Architects/Landscape Architects: Sasaki + Associates
Completed Time: 2008

可持续水资源利用——皇后区植物园新规划

Sustainable Water Resources—New Park Concept for Queens Botanical Garden

翻译 李沐菲

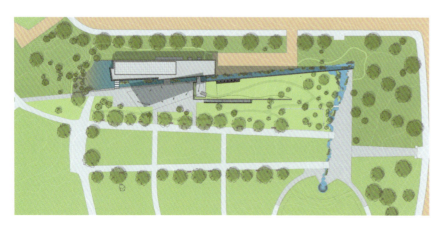

在城市同质性和匿名性的大熔炉里，是否还存在具有独特地域特征的空间？皇后区位于纽约市肯尼迪机场和拉瓜迪亚机场之间，共有142种正式注册语言。皇后区繁华的大街上充斥着各种各样的面孔，民族多样性在此展现得分外明显。皇后区植物园深受当地居民的喜爱，它不仅提供了人们清晨打太极的清净一隅，也是俄罗斯养蜂人、韩国花卉展以及举办浪漫的西班牙式婚礼的理想场所。但是，其年久失修的现状完全抹杀了它在当地人心目中的重要性。

皇后区植物园不仅关注植物的需求，同时也强调人的积极参与。公共研讨会从一开始就邀请当地居民、企业和其他的使用者参与到新公园的设计过程中。水是该设计中的关键主题，作为生命的本源，它将花园中的人、文化和环境特质结合起来。

设计师提出了一项考虑到新规划需求的总体方案。雨水处理系统在土壤污染物管理、新的基础设施、原有花园元素的恢复以及新建水景花园的选址上都发挥着积极的作用。这些水景花园在主题上体现出文化特质,在功能上则充当雨水管理系统的活性元素。

一期资金用于修建一个新的绿色停车场和新的行政大楼。绿色停车场形如人手,"五指"从花园里伸出,充当停车道。在这些停车道上栽种植被,作为雨水滞留带。花园内有一大片草坪采用了一种特殊的土壤基质和地下排水系统,可以在盛大活动期间充当临时停车场。新的行政大楼则完全沉浸在周围北美本土植被营造的美景之中,绿色屋顶和雨水、灰水处理系统生动地体现出绿色建筑原则。

该项目的景观设计和行政大楼设计荣获了美国绿色建筑委员会 LEED 白金设计奖。尤其是其先进的水处理概念,具有明显的优势。水的传统功能将人们在日常生活中水的作用简单化、模糊化,如清洁、运输和处理废物等功能。可持续性水资源管理不仅使该项目美观、技术透明,而且为城市环境做出了杰出的环保贡献。在皇后区植物园中,Atelier Dreiseitl 公司迈出理性的一步,展示出由水而生的环境重要性、文化传统和精神生活之间的动态协同作用。该项目也是堪称全球思维下见微知著的典范。

Our societies are becoming more and more complex. But in the urban homogeneity and anonymity of the melting pot, is there space for true identity? Queens is a New York City borough topped and tailed by New York's two busy airports, JFK and La Guardia. There are 142 officially registered spoken languages in the borough. Ethnic richness is readily visible in the bustling high street, a world market of faces, fruits and foods. Queens Botanical Garden is much beloved by local residents. It offers a quiet corner for morning Tai' Chi, beehives for a Russian beekeeper, facilities for a Korean flower exhibition and a romantic setting for an elaborate Hispanic wedding. Its general state of dilapidation was completely unreflective of its significance to local residents.

Queens Botanical Garden offers encounters not only with the plants of the world but also with its peoples. A public workshop invited residents, local businesses and users to get involved in the design process for the new park concept right from the start. Water was identified as a key theme, which through its essentialness for life, bonds people and cultures and the environmental reality of the Garden itself.

A master plan was developed taking into account new

programming needs. The stormwater management plan is interactive, actively dealing with soil contamination management, new infrastructure, and restoration of existing garden elements and siting of the new water gardens. These water gardens are thematically expressive of cultural identities and functionally active elements of the stormwater management system.

Funding for Phase 1 allowed for construction of a new green parking lot, and a new administration building. The green parking lot has contoured fingers of park, which extend in from the Garden and define the parking lanes. They are naturally planted and act as stormwater retention swales. A large area of meadow has a special soil substrate and under-drainage to accommodate overflow parking for big events. The new administration building melts into a surrounding landscape of native North American plants. A green roof, and rain and grey water systems demonstrate quite conspicuously how principles of green building can be put into practice.

The administration building and landscapes received a LEED Platinum rating, not least because of the advanced water concept. The typical role of water for functions like cleaning, transport and waste disposal reduces the interplay of water in our lives to simplified and imprecise images. Sustainable water resource management combines aesthetic appeal and technological transparency while making a significant "green" space contribution

to our urban environments. At Queens Botanical Garden Atelier Dreiseitl went a logical step further by proposing a design that looks to express dynamic synergies between the environmental importance, cultural traditions and spiritual practices of water. Queens Botanical Garden contains ideas for a neighbourhood garden thinking on a global scale.

All systems
rainwater fed!

GREY WATER SYSTEM

FOUNTAIN / MEANDER / CANAL

SYSTEM I

WATER BASIN

WATER PLAY
WATER FEATURE

SYSTEM II

CENTRAL RAINWATER CISTERN

POND

SYSTEM III

STREAM

SPRING

NATURAL POND

SEASONAL STREAM

WETLANDS

PARKING RUN-OFF

RUN-OFF TREATMENT POND

SYSTEM IV

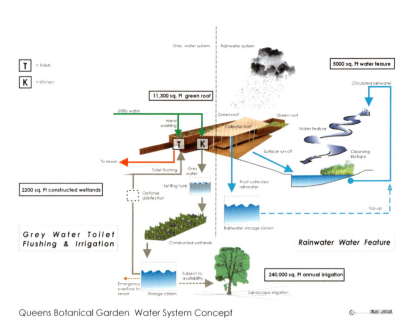

T = Toilets

K = Kitchen

11,300 sq. Ft green roof

5000 sq. Ft water feature

2200 sq. Ft constructed wetlands

Grey Water Toilet
Flushing & Irrigation

Rainwater Water Feature

240,000 sq. Ft annual irrigation

Queens Botanical Garden Water System Concept

The Parking Concept

The parking is envisioned as green fingers which extend into the Garden and are an integrated part of the landscape. Parking areas are shaped by broad areas of planting (a) which provide character and shade and are in themselves planting displays. Areas of overflow parking are surfaced with grass-gravel (b), which allows the parking spaces to team with the adjacent landscape when not in use and facilitates stormwater infiltration. Areas of high use parking are surfaced with permeable paving (c). These surface is hard wearing without the monotony of asphalt, and also allow some infiltration. These paved areas are broken up by areas of planting which extend into the parking bays, emphasizing the overall green character of the parking.

The Parking Drainage

The whole parking surface is composed of permeable to semi-permeable surfaces which allow direct infiltration of rainwater. In cases where there is too much rainfall to be able to all infiltrate directly into the surface (d), it runs into swales (e) - vegetated depressions with high infiltration capacity which run along the planting islands in the parking. These swales have a capacity to handle the majority of rainfalls. In extreme storm events, the swales overflow to the central wetland area (f), ensuring that the parking does not remain water logged.

The planted surfaces of the grass-gravel and swales, and the soil underneath the permeable paving play an important role in treating the surface run-off from the parking. Parked cars drop residues of oil, salt and dirt. This is broken up by micro-bacteria living in the top soil layer.

项目位置：纽约皇后区

客　　户：纽约皇后区植物园

成　　本：1400 万美圆

占地面积：15.4 万平方米

建设时间：2000 年～ 2007 年

景观和水系设计：Atelier Dreiseitl

生态和景观设计：保护设计论坛

建筑设计：BKSK 建筑设计事务所

所获奖项：美国绿色建筑委员会 LEED 白金设计奖

Location: Queens, New York

Client: Queens Botanical Garden, City of New York

Budget: US$14 million

Surface Area: 15.4 ha.

Project Dates: 2000~2007

Landscape and Water Design: Atelier Dreiseitl

Ecology and LA: Conservation Design Forum

Architect: BKSK Architects LLP

Awards: USGBC LEED platinum

制砖厂的华丽转身

The Significant Transformation of a Brick Factory

翻译 刘建明

多伦多是加拿大最大的城市，有"加拿大经济引擎"之称。制砖厂的转型是多伦多后工业时代的标志性景观干预之作。在多学科团队的通力协作下，一块占地约650万平方米的东谷 (Don Valley) 制砖厂和一块后工业时代旧址被改造成为新兴的注重环保的社区中心。该项目由多伦多一个名为"常青树"的非营利组织发起，该组织致力于将城市社区再次完美融入当地生态系统的各种项目中。项目所在地块的当前状态（同时也是本案设计师试图极力改造的）严重制约了该项目所能企及的高度，无奈之下设计师只能退而求其次，尽可能发掘多伦多东部这块"棕地"的潜能。

制砖厂建立于1889年，是加拿大近一个世纪以来国内最大的建筑用砖生产基地。工业采挖形成的大面积锯齿状区域既是历史悠久的工业遗迹，同时也是一块不折不扣的"棕地"。该项目赋予了这个地块全新的景观职能——促进自然、人类与城市之间新理念沟通融合的催化剂。由建筑师Alliance（多伦多）设计的初始规划包括一个占地面积1万多平方米的花园和苗圃、儿童发现乐园、会场与展会设施、溜冰场、一个规划有序的农贸市场、各种社会组织机构的办事机构（包括"常青树"的办事处），以及在当前约650万平方米的湿地和草场范围内的持续开发。

Claude Cormier 的设计理念侧重于将后工业时代的废墟改造成"绿"地。为了跳出日渐为同行业滥用的"可持续性"理念的窠臼，该项目侧重场地的"轨迹运动"理念——水、汽车、电气、火车和野生动物。该项目有针对性地解决了场地多孔性的需保问题，并借此将区域范围内、区域周围以及贯穿区域的可持续景观元素完美地糅合成一套自由流动的体系。

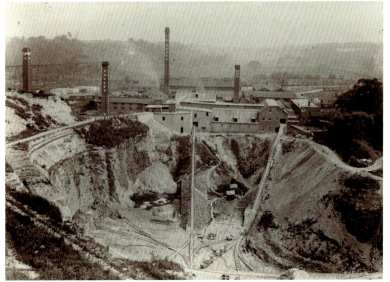

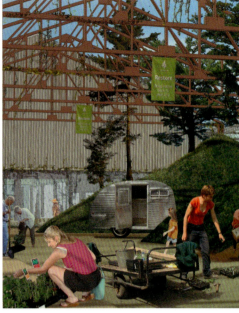

The transformation of an industrial brick factory is one among many iconic post-industrial interventions taking place at a steady pace all over Toronto, Canada's largest city and its economic engine. Through the work of a multidisciplinary team, a quarter of the 40-acre Don Valley Brick Works Park and post-industrial site is being redeveloped as an active, environmentally minded community centre. The project was initiated by Evergreen, a Toronto non-profit organization dedicated to creating programs that reconnect city communities with local natural systems. Its relative success has been limited by the current state of the site, a difficulty that the project seeks to eliminate, instead enhancing the potential of this brown field in East Toronto.

Founded in 1889, the Brick Works was home to one of the country's most significant brick manufacturers for nearly a century. The massive grounds are incised with industrial quarries and are both a heritage site and a brown field. In its new incarnation, the area will serve as a catalyst for new ideas on relationships between nature, individuals, and cities. An initial master plan completed

by Architects Alliance (Toronto) envisages features that include a 110,000 square foot garden and nursery, a children's discovery area, conference and event facilities, skating surfaces, an organic farmers' market, spaces for an array of socially conscious organizations (including the offices of Evergreen), and continued development within the existing 40-acre wetlands and meadows park.

Claude Cormier's conceptual approach focuses on the transformation of this postindustrial ruin into a "green" site. Seeking to transcend the overused term "sustainability", the proposal emphasizes the idea of trajectories of movement through the site – of water, cars, electricity, trains, and wildlife. The plan addresses a need for higher site porosity as a means to create a free-flowing system of sustainable components in, around, and through the area.

项目位置：加拿大安大略省多伦多
客　　户：常青树
占地面积：49 000m²
项目时间：2006 年 ~ 2011 年（在建）
景观设计：Claude Cormier Landscape Architecture + Urban Design
合作伙伴：du Toit Alsop Hillier（多伦多）
　　　　　Diamond+Schmidt Architects Inc.（多伦多）
　　　　　E.R.A. Architects Inc.（多伦多）
　　　　　Surface Architects（英国伦敦）

Location: Toronto, Ontario, Canada

Client: Evergreen

Surface Area: 49,000 m²

Project Dates: 2006~2011 (in progress)

Lead Landscape Architects: Claude Cormier Landscape Architecture + Urban Design

Partners: du Toit Alsop Hillier (Toronto)

Diamond+Schmidt Architects Inc. (Toronto)

E.R.A. Architects Inc. (Toronto)

Surface Architects (London, UK)

景观 LANDSCAPE
设计 DESIGN

景观－设计－艺术

杂志 MAGAZINE

立足本土　　放眼世界

Focusing on the Local　Keeping in View the World

　　《景观设计》杂志自创刊以来一直致力于发掘景观建设中存在的各种环境问题和设计新潮，为中国的城市景观设计、环境规划和城市建设等提供专业化指导，共同推进中国景观设计行业的发展；《景观设计》以其敏锐的视角和专业的办刊风格而成为同类媒体中的先锋杂志，备受业内人士、政府及开发商的推崇。

媒体概要

开本尺寸：225×297mm（16开）　　　语　　种：中英双语

定　　价：48.00元人民币　　　　　　出版周期：双月刊（全年6期）

发行方式：全国发行　　　　　　　　　出版时间：单月20日

地址：大连市甘井子区软件园路80号理工科技园B座1104室（116023）
Address: Room 1104, Block B, No.80 Software Park Road,
Dalian, China (P.C.: 116023)

电话Tel: 86-411-8470 9075　传真Fax: 86-411-8470 9035

大连理工大学出版社·《景观设计》杂志社

征订热线：0411—8470 8943

详情请关注杂志更新内容及www.landscapedesign.net.cn